CREATING MORE EFFECTIVE GRAPHS

D0757407

CREATING MORE EFFECTIVE GRAPHS

NAOMI B. ROBBINS

NBR

Wayne, New Jersey

A JOHN WILEY & SONS, INC., PUBLICATION

Copyright © 2005 by John Wiley & Sons, Inc. All rights reserved.

Published by John Wiley & Sons, Inc., Hoboken, New Jersey.
Published simultaneously in Canada.

No part of this publication may be reproduced, stored in a retrieval system, or transmitted in any form or by any means, electronic, mechanical, photocopying, recording, scanning, or otherwise, except as permitted under Section 107 or 108 of the 1976 United States Copyright Act, without either the prior written permission of the Publisher, or authorization through payment of the appropriate per-copy fee to the Copyright Clearance Center, Inc., 222 Rosewood Drive, Danvers, MA 01923, 978-750-8400, fax 978-646-8600, or on the web at www.copyright.com. Requests to the Publisher for permission should be addressed to the Permissions Department, John Wiley & Sons, Inc., 111 River Street, Hoboken, NJ 07030, (201) 748-6011, fax (201) 748-6008.

Limit of Liability/Disclaimer of Warranty: While the publisher and author have used their best efforts in preparing this book, they make no representations or warranties with respect to the accuracy or completeness of the contents of this book and specifically disclaim any implied warranties of merchantability or fitness for a particular purpose. No warranty may be created or extended by sales representatives or written sales materials. The advice and strategies contained herein may not be suitable for your situation. You should consult with a professional where appropriate. Neither the publisher nor author shall be liable for any loss of profit or any other commercial damages, including but not limited to special, incidental, consequential, or other damages.

For general information on our other products and services please contact our Customer Care Department within the U.S. at 877-762-2974, outside the U.S. at 317-572-3993 or fax 317-572-4002.

Wiley also publishes its books in a variety of electronic formats. Some content that appears in print, however, may not be available in electronic format.

Library of Congress Cataloging-in-Publication Data:

Robbins, Naomi B.
Creating more effective graphs / Naomi B. Robbins.
p. cm.
Includes bibliographical references and index.
ISBN 0-471-27402-X (pbk.)
1. Statistics–Graphic methods. I. Title.

HA31.R535 2005
519.5′02′1–dc22

2004049163

Printed in the United States of America

10 9 8 7 6 5 4

To Ed, Joyce, and Rich

CONTENTS

PREFACE

The idea for this book came while reading *The Elements of Graphing Data* by William S. Cleveland. Cleveland addresses his book to people in science and technology. As I read it, I kept thinking that his ideas deserved a wider audience: that people in the business world, the financial world, the world of nonprofits, and many other groups would benefit from his ideas and methods. *Creating More Effective Graphs* addresses this need.

Goals

This book is intended to be quick and easy to read. It is not intended to be a substitute for the classic books in the field: Cleveland's *The Elements of Graphing Data* and Edward Tufte's *The Visual Display of Quantitative Information*. Rather, it is hoped that it will serve as an introduction to the subject, encouraging readers to study the classics. The number of confusing, poor graphs one sees is staggering. However, at a recent visit to a business school library, I was delighted to see how many more clear, accurate, well-designed graphs appear in journals currently than they did a decade ago. The credit for this improvement goes to the influence of these authors.

Audience

This book offers guidance to readers with a wide range of backgrounds. As a result, at times I review some high school math, as I do in Chapter 4 when reviewing logarithms. Some readers will appreciate this review. At other times I include more complex technical details for the more technical readers. Nontechnical readers can feel comfortable skipping these sections without worrying about being able to understand later parts of the book. This book is a collection of independent examples and advice; it is not like a course, where material builds on earlier sections, keeping you lost forever if you do not understand a fundamental concept. So please skip the paragraphs that are too easy or too hard for you and benefit from the rest. (Note that Chapter 3 provides basic information useful when reading later chapters.)

Terminology

Unfortunately, there is little standard terminology in the literature on data graphs. Words such as *chart*, *graph*, and *plot* are often used interchangeably. Some authors have given precise meanings to these words and have used them consistently. The problem is that they then no longer use the term of the creator of the graph. For example, John Tukey (1977) introduced a box and whisker plot in his 1977 book. For over 25 years, statisticians have called them box plots or box and whisker plots. I am unwilling to call them box graphs for the sake of consistent terminology. So my choice has been to continue with the name that the originator of the technique used. The term *graph* in this book always means data graph, as opposed to the nodes and edges of a graph in the branch of mathematics called graph theory.

Examples

You will notice that a disproportionate number of examples come from the world of museums. There are a number of reasons for this. I have found that the museum community is very interested in the accurate display of information and wants to display information effectively. Their data are less likely to be private, so they are more willing to allow me to share their data. We all understand what we mean by their terms, so I don't have to waste space defining the concepts behind the data. Finally, since many of us go to museums, we can relate to the examples. If I used examples from the pharmaceutical industry, people in nonprofits or marketing would feel that the examples do not apply to them.

Wherever possible, the figures are real figures with real data. In some cases, using real data would violate confidentiality, so labels or data have been changed. In other cases, I have reproduced real problems that are frequently seen with simulated data.

How to Read

You can start at the beginning and read through the text or use the book as a reference to look up the topics you need. Either way, I recommend reading Chapter 3 before later chapters, since it helps to understand why some presentations work and some do not. Reading the entire book is not a big investment of time.

Chapter 1 explains what we mean by an effective graph.

Chapter 2 shows problems with many of the common graphs that are ubiquitous today.

Chapter 3 briefly discusses the tasks we perform when we decode information from a graph, and which of these tasks we perform well.

Chapters 4 and 5 present methods. Chapter 4 describes graphs with one or two variables that are more effective than those of Chapter 2. Graphs with more than two variables are described in Chapter 5.

Chapters 6 and 7 contain principles. **Chapter 6** presents general principles for creating effective graphs. Principles for choosing scales appear in **Chapter 7**.

Chapter 8 applies what we have learned by looking at *before* and *after* examples.

Chapter 9 contains some comments about software.

Chapter 10 includes questions and answers.

Appendix A is a checklist of graph defects.

Appendix B lists all figures and their sources.

The book contains examples of both good and bad graphs. It should be clear from reading the book which is which. However, to prevent readers who just skim the book from thinking that I recommend the bad graphs, the icon you see on the margin is placed near some graphs. The absence of an icon does not necessarily mean that the graph form or the editing choices are recommended.

Tools Needed

All of the figures drawn for this book were produced on home equipment and print well on an inkjet printer. The one luxury was powerful software, S-Plus, available from Insightful

Corporation. Equally powerful open-source software, R, is freely available to download. It does not take a large budget to improve your graphs.

Recognizing that many people make graphs with Microsoft Excel, much of the advice applies to Excel users, and I provide some tips for Excel users in Chapter 9.

Acknowledgments

You will soon become aware that every chapter in the book has been influenced by the work of William Cleveland. I cannot thank Bill enough for his generosity in sharing his ideas, encouraging me in this field, and allowing me to quote his materials.

As deadlines approached, Linda Clark helped to produce some of the figures. The list of figures in Appendix B credits her for the ones she drew. Paul Murrell provided some code for Figure 8.9 before the documentation for the features used was available. Kenneth Klein wrote a macro to allow Excel users to draw dot plots. Marc Tracey provided me with information on Illustrator for Chapter 9.

Thanks to all who allowed me to use their unpublished data or figures: Edith Flaster, the Monterey Bay Aquarium, Beverly Serrell, the St. Louis Science Center, Marc Tracey, private clients, and anonymous contributors. Thanks also to the Center for the Study of Philanthropy, the Graduate Center, CUNY, for its contribution.

This project would never have gotten off the ground without the encouragement of Steve Quigley, Executive Editor at Wiley. It benefited throughout from the top-notch skills of the Wiley team, from the copyeditor to the compositor. Special thanks go to Angioline Loredo, Associate Managing Editor. It was a pleasure to work with her.

The book benefited greatly from reviewers, who gave generously of their time to tell me what they liked and didn't like, and what they understood and didn't understand: Edith Flaster, Anne Freeny, Bert Gunter, Amy Juviler, Nancy Klujber, Melanie Meharchand, Beth Lisberg Najberg, Edward Robbins, Joyce Robbins, Richard Robbins, Ervin Schoenblum, and an anonymous reviewer from Wiley. I appreciate their candid remarks.

1 Introduction

In today's world we are overwhelmed with data. Graphs can be incredibly powerful tools in creating order from the chaos of numbers. A basic knowledge of graphing techniques is needed to ensure that data are presented effectively.

This book will teach you:

- How to make clear and accurate graphs that improve understanding of data
- How to avoid common problems that cause graphs to be ineffective, confusing, or even misleading
- When to use new graphing techniques to simplify complex data presentation
- How to be more critical and analytical when viewing graphs

You will learn guidelines and principles for creating effective graphs. These principles are to numbers what the rules of grammar are to words. Not many people today write without a computer. But we all know that being skilled at using word processors does not make us great writers. Similarly, learning how to use a software package to make graphs is not enough to become an effective communicator of numerical data.

Creating More Effective Graphs, by Naomi B. Robbins
ISBN 0-471-27402-X Copyright © 2005 John Wiley & Sons, Inc.

Fig. 1.1 Similar Pie Wedges

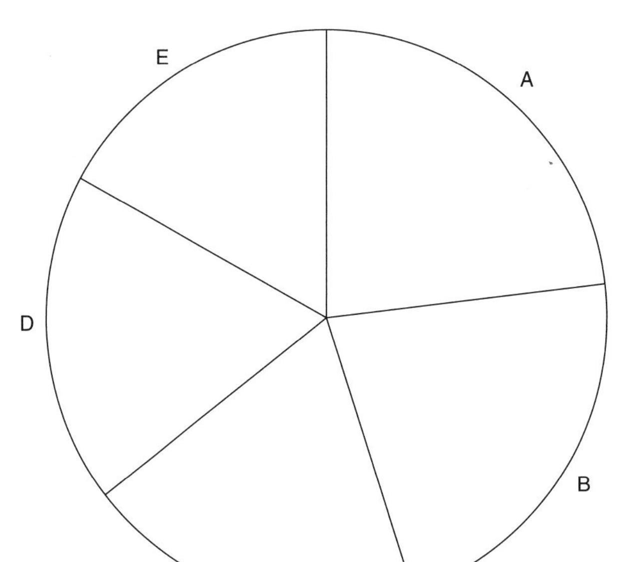

Whether you draw graphs manually or with a computer, and regardless of the software package you use, increasing your knowledge of the principles of effective graphs will greatly improve your work.

1.1 WHAT WE MEAN BY AN EFFECTIVE GRAPH

We begin by discussing what we mean by an *effective graph*. Figure 1.1 is the familiar *pie chart*. This one has five wedges labeled A through E. Study this chart and try to place the wedges in size order from largest to smallest. Use a pencil and paper to write down your results.

Fig. 1.2 Similar Pie Wedges: Dot Plot

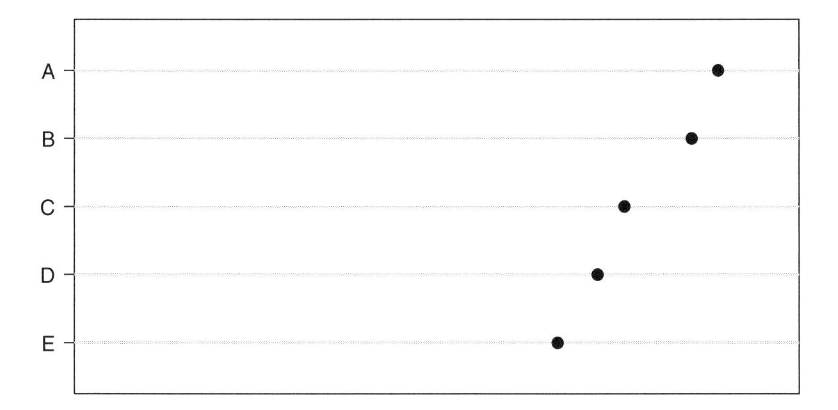

The same data are plotted in Figure 1.2, this time using a *dot plot*. Once again take out paper and pencil and order the size of categories A through E.[1] Most of you probably had a lot of trouble placing the wedges of the pie chart in size order. Some of you may have found this to be quite frustrating, even though a few of you probably had no trouble at all. But even those of you who could order the wedges of the pie chart easily must admit that this task is much easier using a dot plot. Cleveland (1984) introduced dot plots to take advantage of the results of experiments on human perception and the decoding of graphical information. In Chapter 3 I discuss this topic briefly.

[1] I have left out the tick marks and labels since a reviewer suggested that it was an unfair comparison to show tick labels on the dot plot and no labels on the pie chart. Tick mark labels help you estimate the values, which is not the task that you were asked to do.

Now we're ready to define what we mean by an *effective graph*. One graph is more effective than another if its quantitative information can be decoded more quickly or more easily by most observers. Here, Figure 1.2, the dot plot is more effective than Figure 1.1, the pie chart.

This definition of effectiveness assumes that the reason we draw graphs is to communicate information—but there are actually many other reasons to draw graphs. Figures help to keep an audience attentive during presentations. A graph may make a page in a document more attractive and inviting by adding variety and avoiding a totally gray page of text, thereby increasing readership. This book and our definition of effectiveness do not address these other reasons except in the questions and answers in Chapter 10. There you will see that we should communicate clearly and accurately even if our primary purpose for including a graph is one of the other reasons that I have listed.

Fig. 1.3 Similar Pie Wedges: Table

A	23.0
B	22.0
C	19.5
D	18.5
E	17.0
Total	100.0

Sometimes graphs do not provide the best solution for presenting data. A table provides another way to show the data in Figures 1.1 and 1.2. The table in Figure 1.3 not only shows the exact values of the categories, but also shows the total. Tables are preferable to graphs for many small data sets.

1.2 GENERAL COMMENTS

1.2.1 Captions

Cleveland (1994) says that a graph should include a caption that draws attention to important features of the data and describes conclusions that are drawn from the data. Why, then, although we use identifying figure headings in this book, did we not use captions? It is because the style of the book is to make you think about graphs and answer questions before you read the material that normally is contained in a caption.

1.2.2 The Data We Plot

The information to be plotted may be quantitative or categorical. Quantitative variables have numerical values (e.g., heights, measurements, salaries). Categorical variables have labels as values (e.g., gender, country, occupation).

An early step in any statistical analysis, including the presentation of data, is to check and clean the data. Are there typos? Do the numbers make sense? Why are a few data points far from the others? Are there inconsistencies, such as a birth date later than the corresponding death date? The advice in this book applies *after* the data have been checked. We assume that the numbers are correct and discuss how to display them effectively. However, another value of graphing is that graphs are a useful tool in helping to check data. We will see several examples where graphs have helped to identify incorrect data. Best (2001) discusses how numbers get distorted and take on lives of their own.

SUMMARY

One graph is more effective at communicating quantitative information than another if most readers can decode the quantitative information from it more quickly or more easily. Tables are useful for displaying small data sets.

2 Limitations of Some Common Charts and Graphs

Fig. 2.1 Structured Data Set

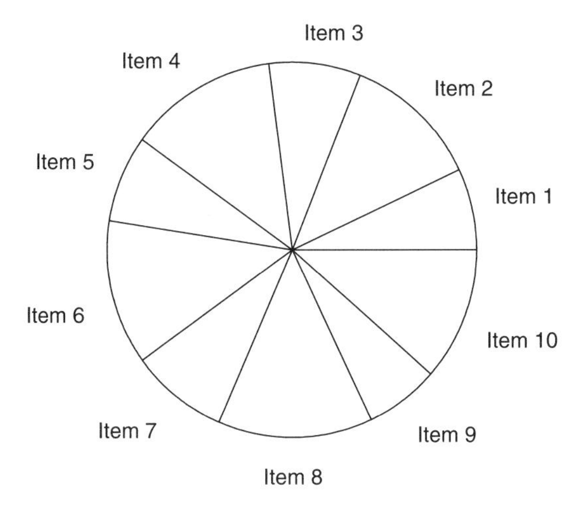

Creating More Effective Graphs, by Naomi B. Robbins
ISBN 0-471-27402-X Copyright © 2005 John Wiley & Sons, Inc.

2.1 PIE CHARTS

The pie chart in Figure 2.1 shows 10 wedges, which I have labeled items 1 through 10. Many authors suggest ordering pie charts from largest to smallest unless there is a natural ordering[1] of the data, and our labeling here gives a natural ordering. I would like you to study the data and learn from them what you can.

[1] Jan., Feb., and Mar. are ordered categories; so are low, medium, and high and first, second, and third. Apples, oranges, and bananas are categorical data that do not have a natural ordering.

Fig. 2.2 Structured Data Set: Dot Plot

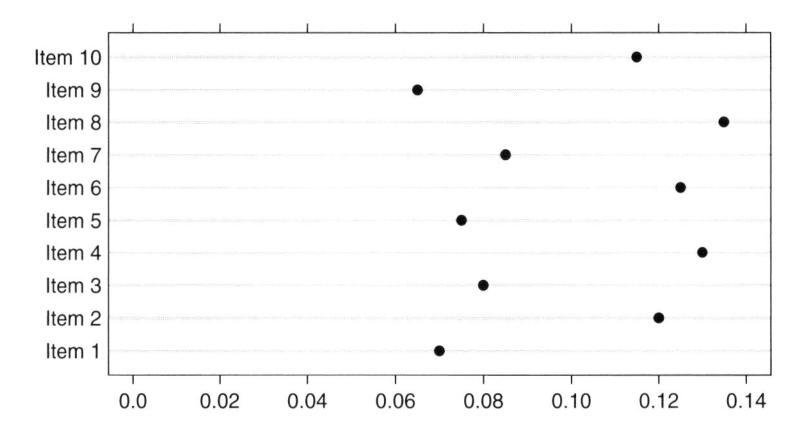

Some of you probably noticed in Figure 2.1 that the areas of the wedges labeled with odd numbers were smaller than those with even numbers. Once again we show the same data in a dot plot. Study Figure 2.2 and see what you can learn about the data.

This is a very structured set of data. The values of the items with even-numbered labels are an exact offset of those with odd-numbered labels. Those with odd-numbered labels are centered on 0.075, and each item with an even-numbered label is exactly 0.05 bigger than the preceding one with an odd number. I challenge you to see this in Figure 2.1, but it is clear in Figure 2.2.

William Cleveland and Edward Tufte are authors who have written excellent books on graphing data. Becker and Cleveland (1996, p. 50) say: ''Pie charts have severe perceptual problems. Experiments in graphical perception have shown that compared with dot charts, they convey information far less reliably. But if you want to display some data, and perceiving the information is not so important, then a pie chart is fine.'' Some readers think that this quote must be tongue-in-cheek. Perhaps they are saying that if you want to present a few numbers in a pie chart to decorate the page rather than communicate clearly, no real harm is done.

Tufte (1983, p. 178) prefers a table to a "dumb pie chart; the only worse design than a pie chart is several of them, for then the viewer is asked to compare quantities located in spatial disarray both within and between pies. . . . Given their low data-density and failure to order numbers along a visual dimension, pie charts should never be used."

Fig. 2.3 Three-Dimensional Pie Data

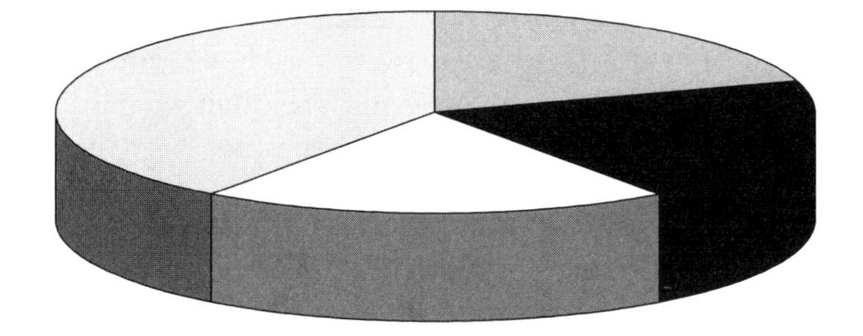

2.2 CHARTS WITH A THREE-DIMENSIONAL EFFECT

Tufte said that the only design worse than a pie chart was several of them, but I consider the design shown in Figure 2.3 to be worse than a simple pie chart: It is a *three-dimensional pie chart*. You know that the four wedges add up to 100%. Take your pencil and paper and write down four numbers adding to 100 that represent the percentage of each wedge.

Fig. 2.4 Three-Dimensional Pie Data: Two-Dimensional Bar Chart

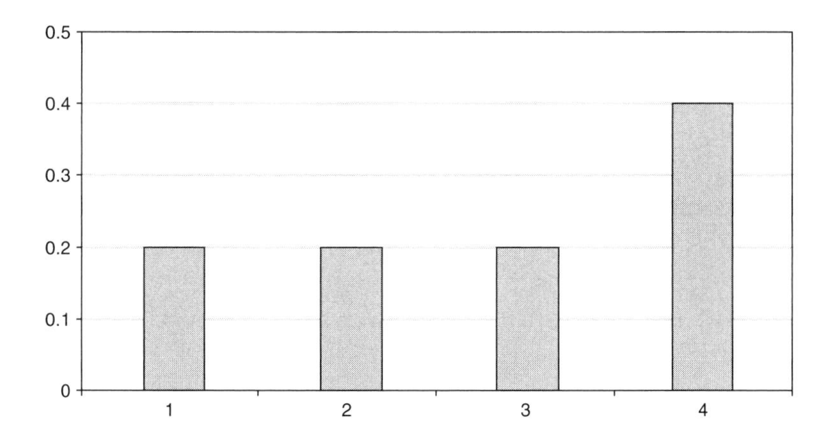

In Figure 2.4 I show the same data on a simple *two-dimensional bar chart*. Now you see clearly the percentage of each category. Are the results the same as those you read from Figure 2.3?

Fig. 2.5 Three-Dimensional Pie Data: Two-Dimensional Pie Chart

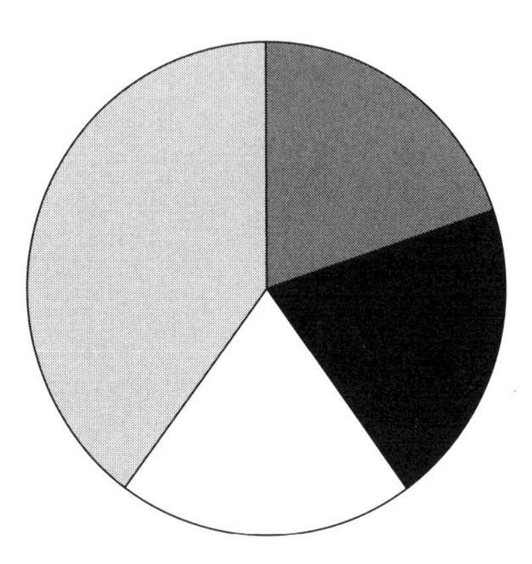

Fig. 2.6 Three-Dimensional Bar Data

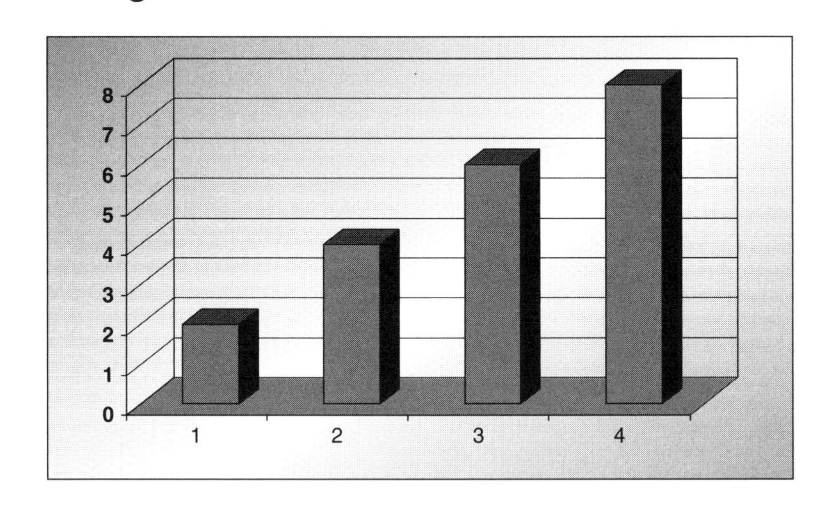

We have already seen the limitations of pie charts. But the *two-dimensional pie chart* in Figure 2.5 is certainly an improvement over the three-dimensional version.

Three-dimensional bar charts[2] such as that in Figure 2.6 are ubiquitous. Take your pencil and paper and write down the heights of the bars.

[2] A real three-dimensional bar chart displays three variables; charts with two variables, as in Figure 2.6, are frequently called *pseudo-three-dimensional bar charts*. Common terminology is used here since bar charts with three variables are not discussed.

Fig. 2.7 Three-Dimensional Bar Data: Two-Dimensional Bar Chart

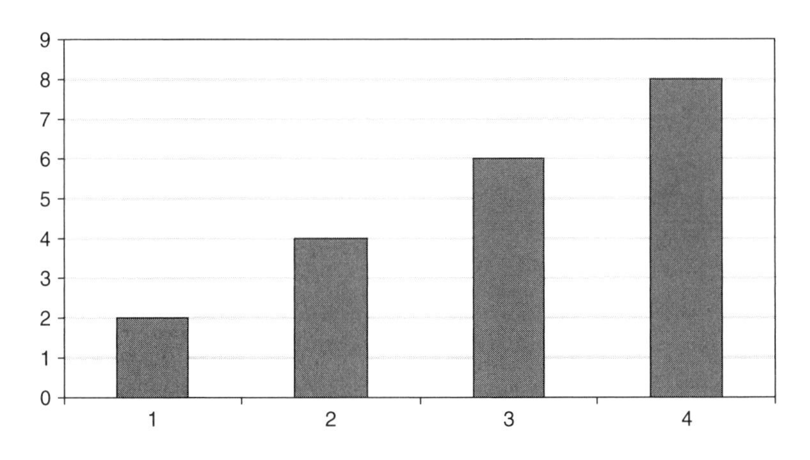

Fig. 2.8 Three-Dimensional Bar Data: Excel

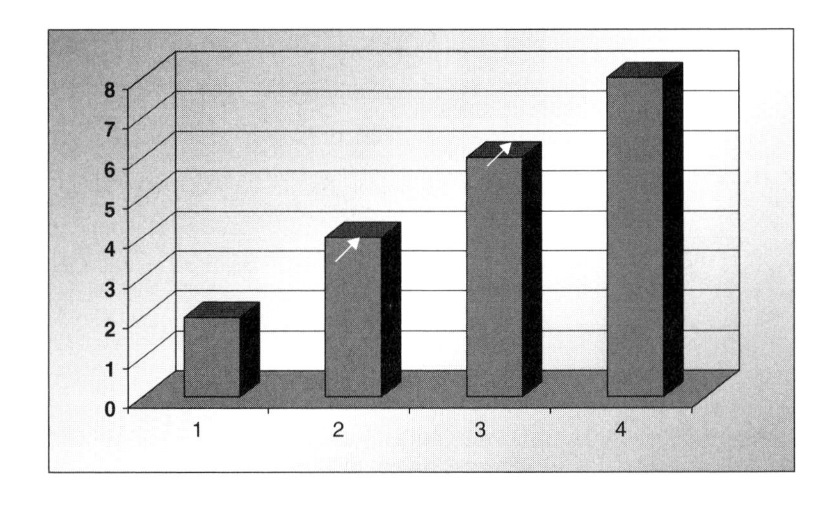

Again, I show the same data in a *two-dimensional bar chart* (Figure 2.7). Do you still agree with the numbers you wrote down?

The problem with three-dimensional bar charts such as Figure 2.8 is that virtually no one knows how to read them. Do you read from the front of the bar, as shown by the arrow on bar 2, or from the back of the bar, as shown by the arrow on bar 3? Figure 2.8 was drawn using Excel 2000. We see in this case that neither supposition is correct. I assume that we are supposed to visualize a plane with a height of 2, and that the top of the first bar is supposed to look tangential to that plane. It doesn't work for me. Does it work for you?

Fig. 2.9 Three-Dimensional Bar Data: PowerPoint

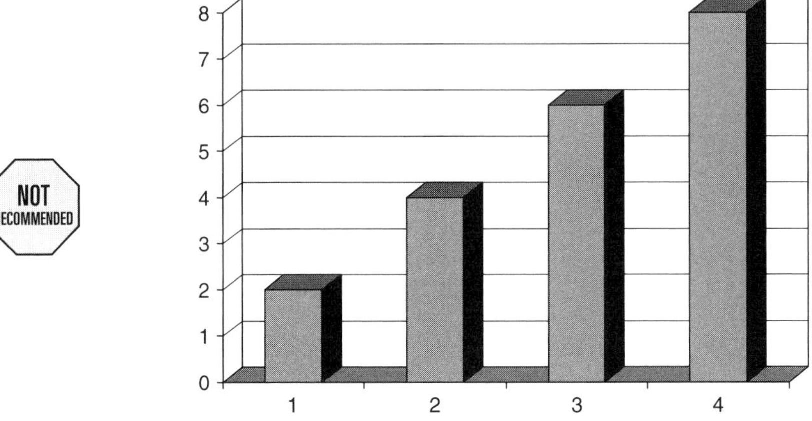

Fig. 2.10 Three-Dimensional Bar Data: Presentations & Charts

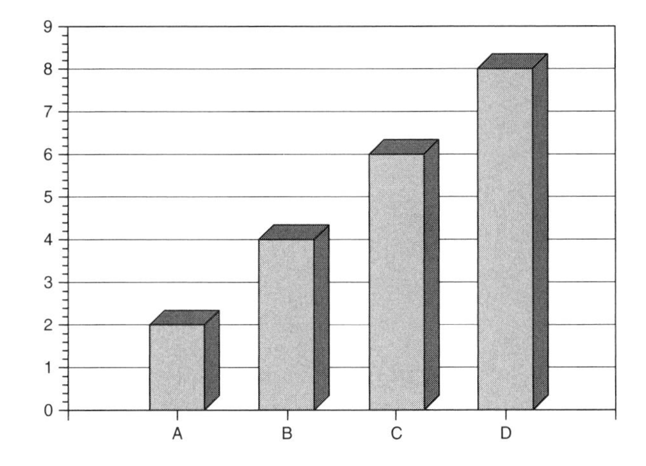

This page shows the same data drawn using two other programs. Figure 2.9 was prepared with Microsoft Graph in PowerPoint; Figure 2.10, with an inexpensive charting program called Presentations & Charts. In the PowerPoint case we read the bars from the back, and in Presentations & Charts, from the front.

These examples demonstrate that the way to read three-dimensional bar charts depends on the software used to create them. But the reader rarely knows what software was used so has little hope of reading them correctly without the values printed. Even PowerPoint and Excel, two programs that come packaged together in the same suite, use different algorithms to plot their graphs. Therefore, you should never use a three-dimensional bar chart for two variables. A properly drawn two-dimensional chart shows the same information more effectively and avoids misinterpretation.

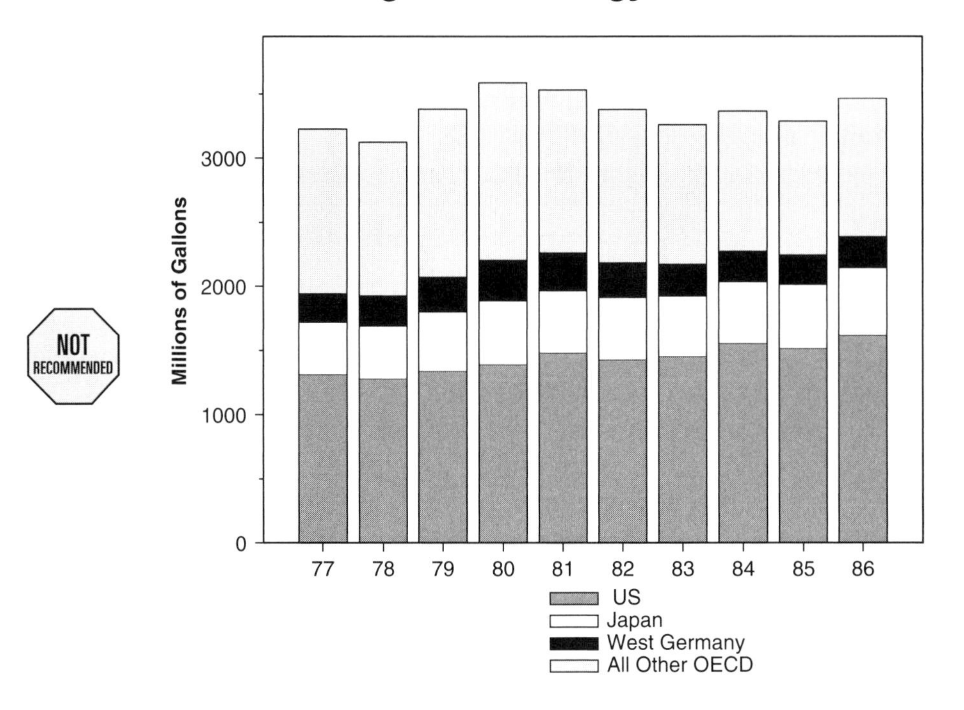
Fig. 2.11 Energy Data
NOT
RECOMMENDED
Millions of Gallons
0
1000
2000
3000
77
78
79
80
81
82
83
84
85
86
US
Japan
West Germany
All Other OECD

2.3 BAR CHARTS: STACKED AND GROUPED

Another common graph form is a *stacked bar chart*. Figure 2.11 shows petroleum stocks from 1977 to 1986 in millions of gallons for the United States, Japan, West Germany, and all other countries of the Organisation for Economic Co-operation and Development (U.S. Dept. Energy, 1986). You probably read the values for the United States and the totals quite accurately. Study the chart and see what you can discover about the other countries.

Fig. 2.12 Energy Data: All Other OECD

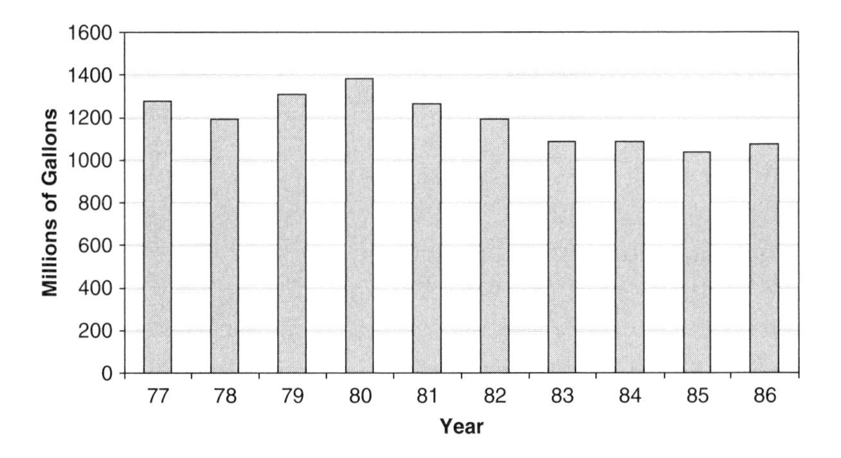

Did you notice in Figure 2.11 that the values for ‘‘all other OECD’’ generally tend to decrease over time? You probably didn’t. As we shall see in Chapter 3, it is very difficult to judge lengths that do not have a common baseline.

Fig. 2.13 Energy Data: Grouped Bar Chart

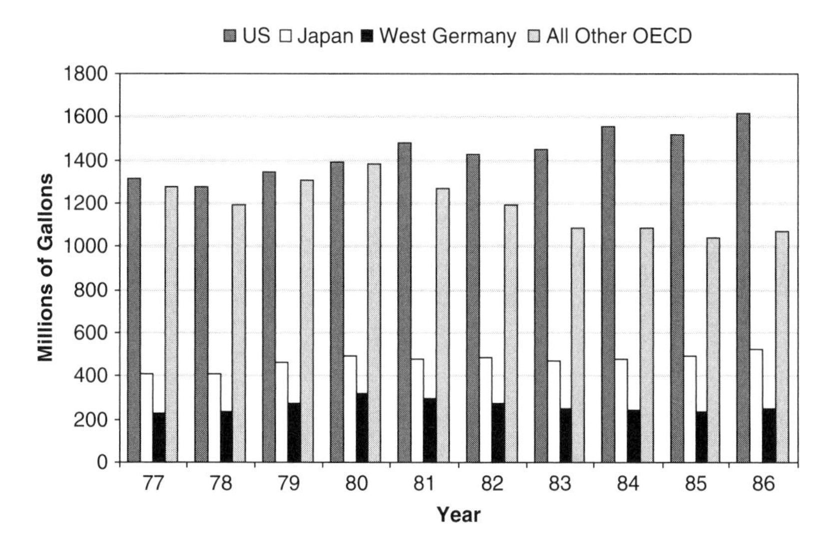

The bars in *grouped bar charts* do have a common baseline. However, a grouped bar chart such as Figure 2.13 becomes difficult to read with even a few groups. It is difficult to follow the trend for a given group such as Japan because the data for the other groups fall between the consecutive values for Japan. Reordering the shadings helps to make the groups distinguishable. The pattern of the "all other OECD" group is certainly clearer than in the stacked bar chart. However, trellis displays, which are discussed in Chapter 5, are far clearer than is a grouped bar chart.

Fig. 2.14 Playfair's Balance-of-Trade Data

2.4 DIFFERENCE BETWEEN CURVES

Most of the graph forms that have been used until recently were introduced by William Playfair in the late eighteenth and early nineteenth centuries. Figure 2.14 uses Playfair's data (Playfair, 1786) to show exports from England and imports to England in trade with the East Indies. We're interested in the balance of trade, which is the difference between exports and imports. We see that the difference is about 0.4 minus 0.2 or 0.2 in 1700, and then it increases for awhile. I'd like you to continue sketching the difference.

Fig. 2.15 Playfair's Balance-of-Trade Data: Imports Minus Exports

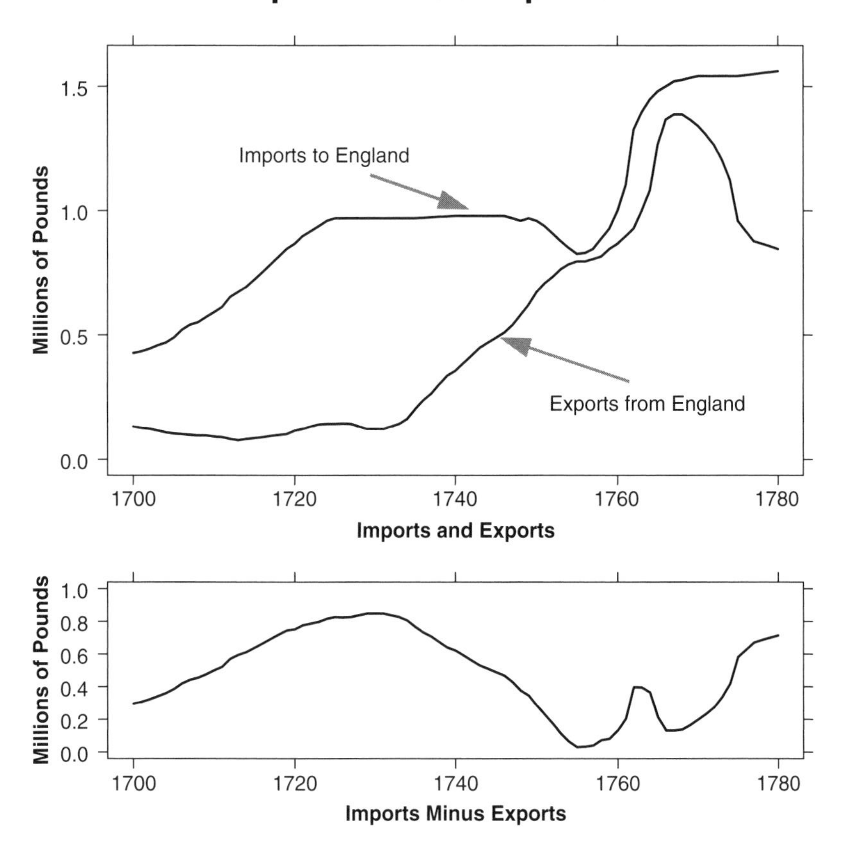

Did you notice the hump after 1760? We miss it because our eyes look at the closest point rather than the vertical distance.

It is important to remember to plot the variable of interest. If interested in the balance of trade, plot the difference rather than just the imports and exports. If we have *before* and *after* data and are interested in improvement, we plot the improvement, not just the *before* and *after* data.

Fig. 2.16 Difference between Curves

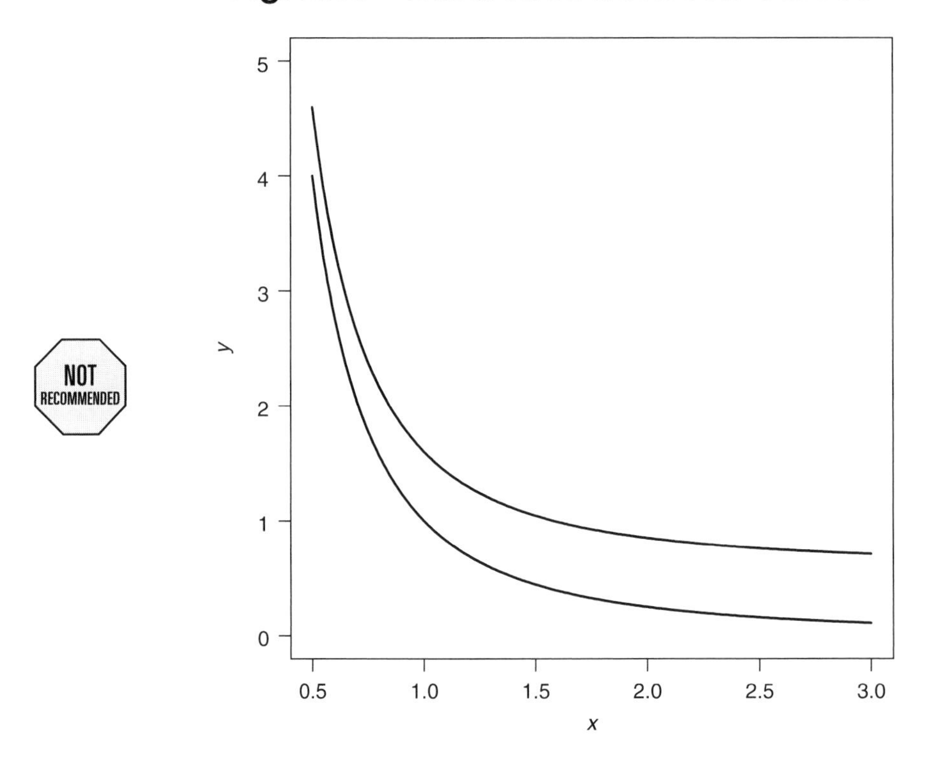

Look at the two curves in Figure 2.16. For what values of x are they closest together and also, farthest apart?

Same all the way!

Fig. 2.17 Ownership of Government Securities

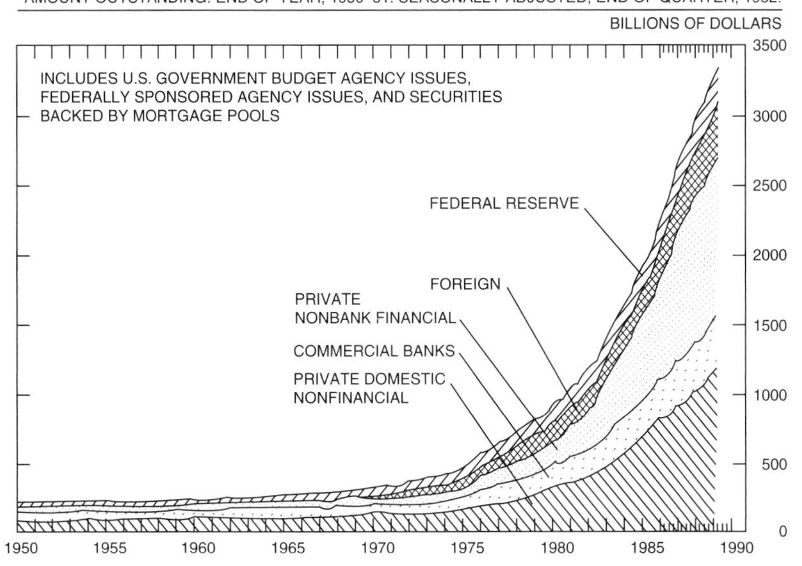
OWNERSHIP OF U.S. GOVERNMENT SECURITIES
AMOUNT OUTSTANDING: END OF YEAR, 1950–51: SEASONALLY ADJUSTED, END OF QUARTER, 1952.
BILLIONS OF DOLLARS
INCLUDES U.S. GOVERNMENT BUDGET AGENCY ISSUES, FEDERALLY SPONSORED AGENCY ISSUES, AND SECURITIES BACKED BY MORTGAGE POOLS
FEDERAL RESERVE
FOREIGN
PRIVATE NONBANK FINANCIAL
COMMERCIAL BANKS
PRIVATE DOMESTIC NONFINANCIAL
3500
3000
2500
2000
1500
1000
500
0
1950
1955
1960
1965
1970
1975
1980
1985
1990

NOT
RECOMMENDED

The last question appeared to be easy, but actually the two curves differ by a constant amount. The curves plotted are $y_1 = 1/x^2$ and $y_2 = y_1 + 0.6$, so that one curve is always exactly 0.6 higher than the other.

The last few charts have taught me not to trust my judgment when viewing charts such as the ownership of U.S. government securities [Board of Governors of the Federal Reserve System (U.S.), 1989]. If interested in how a specific group, say commercial banks, changed over time, I would perform the subtraction and plot that group over time as we did with the exports and imports in the Playfair example.

Fig. 2.18 Playfair's Population of Cities

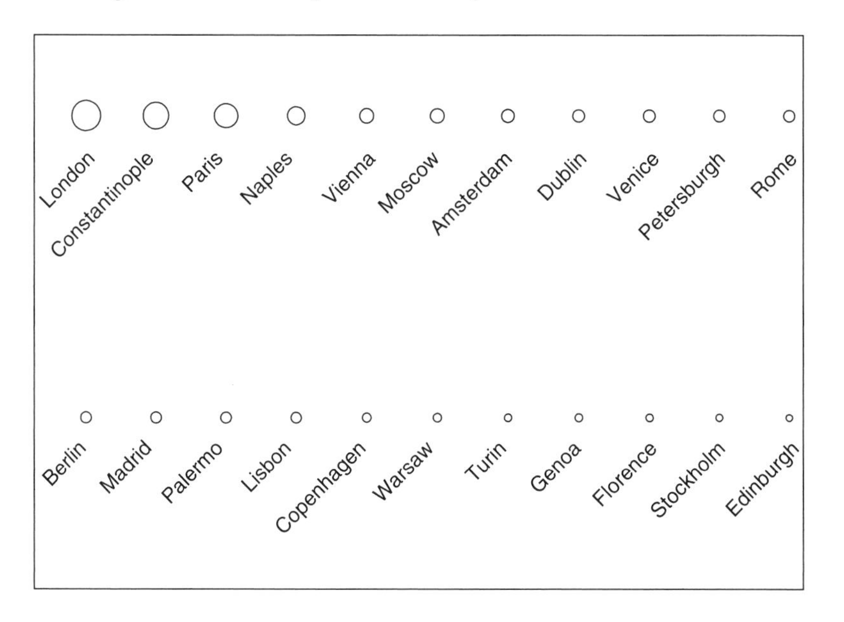

2.5 BUBBLE PLOTS

Figure 2.18 is another plot based on data of Playfair (Playfair, 1801); this one shows the population of cities at the end of the eighteenth century. Since we don't distinguish small differences in area well, it is difficult to place these cities in size order.

Fig. 2.19 Population of Cities: Dot Plot

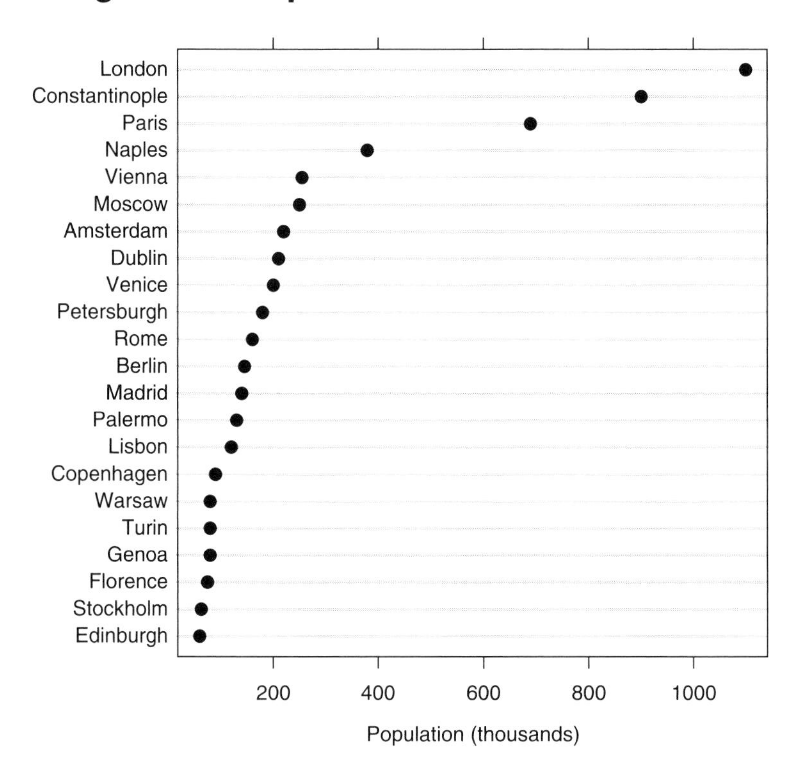

Once again we replot the data using a dot plot. It is now easy to order areas that appeared the same in the bubble chart. It is also easier to determine the population of these cities.

SUMMARY

Some graph forms, including pie charts, bubble charts, and stacked bar charts, are difficult to read accurately, and they hide the structure of the data. In Chapters 4 and 5 we look at more effective graphs that can be used in their place. Never use three-dimensional charts for two-dimensional data. Virtually no one can read them.

3 Human Perception and Our Ability to Decode Graphs

In Chapter 2 we saw that some common graphs do not communicate numerical information effectively. We also discovered other graphs that clearly communicate the information and the patterns of the data. In this chapter we examine briefly the tasks required to decode the information in a graph. Cleveland and McGill (1984) ran experiments to determine which of these tasks we do most accurately. This knowledge helps us to understand why some graphs work and others don't. We first list in alphabetical order 10 judgments we make when decoding quantitative information from graphs, describe each briefly, then order them by our ability to perform them accurately. In some cases the descriptions of two tasks with similar properties (e.g., area and volume) appear together.

Creating More Effective Graphs, by Naomi B. Robbins
ISBN 0-471-27402-X Copyright © 2005 John Wiley & Sons, Inc.

3.1 ELEMENTARY GRAPHICAL PERCEPTION TASKS

Angle

Area

Color hue

Color saturation

Density

Length

Position along a common scale

Position along identical, nonaligned scales

Slope

Volume

Fig. 3.1 Angle Judgments

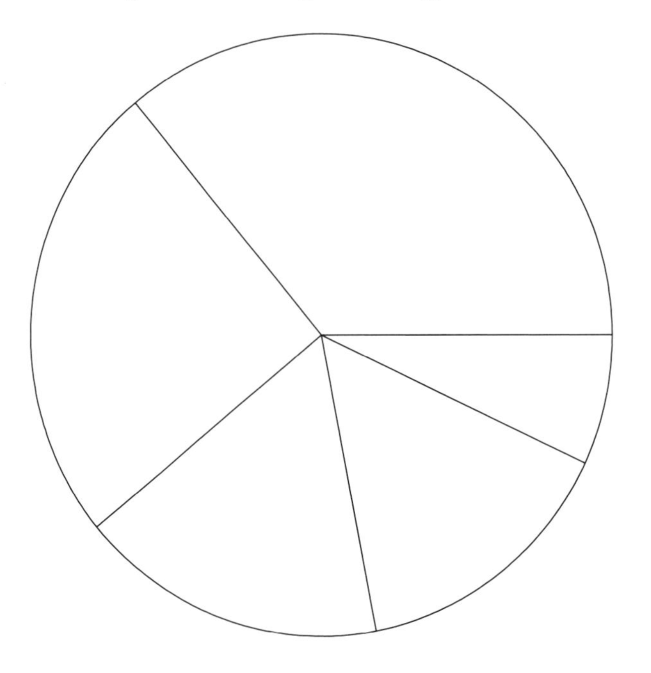

We make angle judgments when we read a pie chart, but we don't judge angles very well. These judgments are biased[1]; we underestimate *acute angles* (angles less than 90°) and overestimate *obtuse angles* (angles greater than 90°). Also, angles with *horizontal bisectors* (when the line dividing the angle in two is horizontal) appear larger than angles with vertical bisectors.

---

[1] *Biased* means that we consistently under- or overestimate the true value.

Fig. 3.2 Area and Volume Judgments

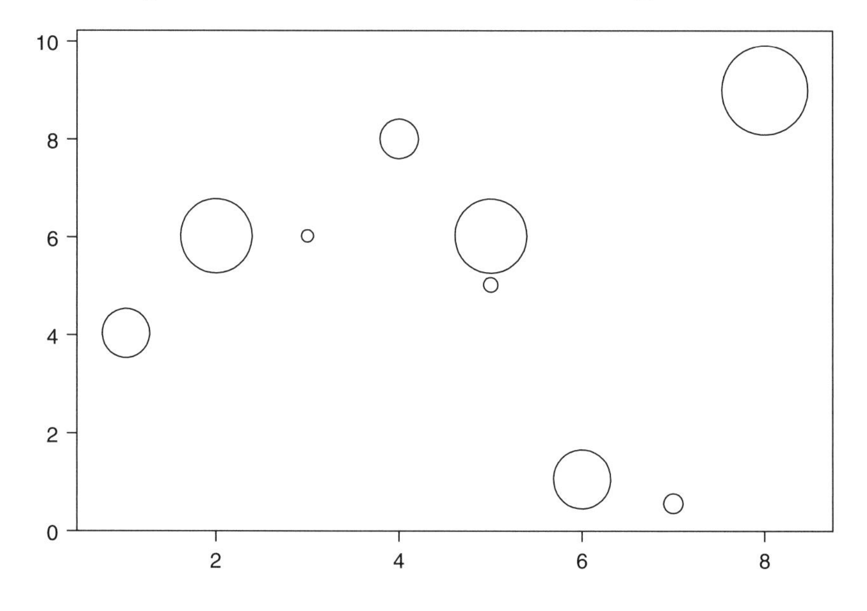

The circles on Figure 3.2 show three variables by the horizontal position of the center of the circle, the vertical position, and the area of the circle. For example, the horizontal axis could be the months that you have held a security, the vertical could be the price you paid in thousands of dollars, and the areas of the circles could represent your gain. These charts are often called *bubble plots*.

Area judgments are also biased. They are much less accurate than length and position judgments. Volume judgments are even more biased. Stevens (1975) presents the following law: Let x be the magnitude of an attribute of an object, such as its length or area. According to *Stevens' law*, the perceived scale is proportional to x^β, where β has been determined by experimentation to range generally from 0.9 to 1.1 for length, 0.6 to 0.9 for area, and 0.5 to 0.8 for volume.

When $\beta = 1$, $x^\beta = x$, and when $\beta < 1$, $x^\beta < x$. Since the β for area is less than 1, we perceive areas to be smaller than they really are. This bias is more pronounced with volumes. The range of beta for lengths includes 1, so we perceive lengths more accurately than areas or volumes.

Fig. 3.3 Color Hue, Saturation, and Density

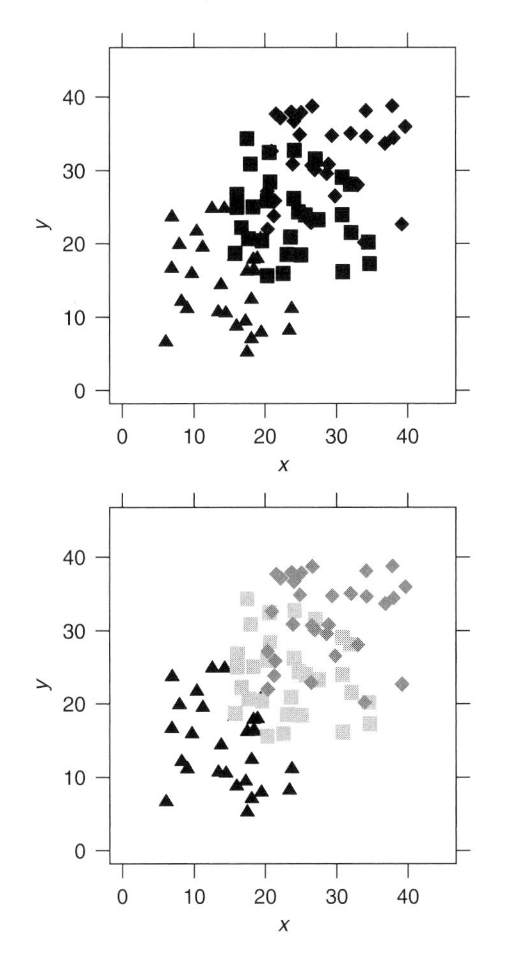

Color coding (*hue*[2]) is very effective for distinguishing data from various groups. For example, suppose that the triangles in Figure 3.3 represent data for England; the squares, France; and the diamonds, Italy. It is difficult visually to separate the three groups in the top plot. Varying density or saturation can also be used to distinguish groups of data. Using different densities in the bottom plot makes this task easier, even though we are limited to shades of gray. If we use just one hue, we can rank by saturation or density to show levels of a quantitative variable. Weather maps often vary shades of red and blue to show temperature. But color coding with different hues does not work well for showing numerical information since we don't perceive an ordering to red, green, blue, and other hues.

[2] *Hue* is the technical term for what we call *color* (red, yellow, blue, etc.). *Saturation* refers to the intensity of the color. As saturation increases, the color becomes purer; as saturation decreases, the color becomes more gray. *Density* refers to the shading or amount of black.

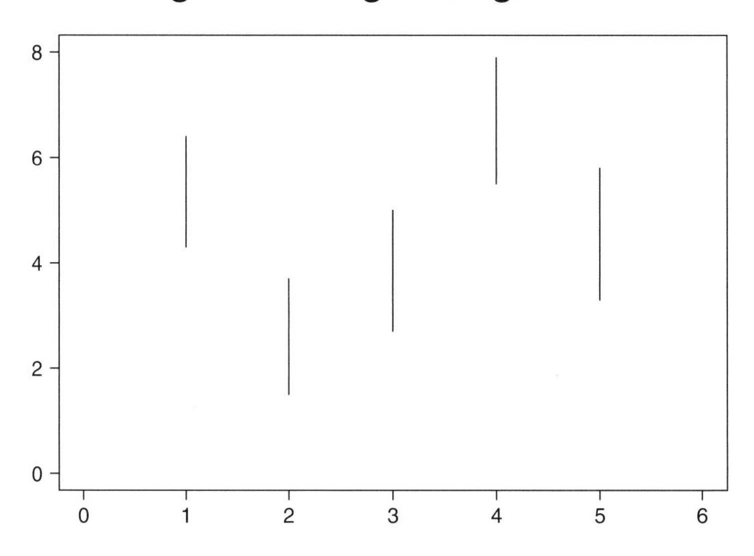
Fig. 3.4 Length Judgments
8
6
4
2
0
0
1
2
3
4
5
6

We all know what length means, but can you order the lengths of the five line segments in Figure 3.4? We learned from Stevens' law that we judge lengths more accurately than areas or volumes, but judging lengths is still not easy.

To detect a difference in length between two line segments, we need a fixed percentage increase in the length. For example, if one line is 99 inches and the other is 100 inches, it will be much more difficult to distinguish this 1-inch difference than if one line is 1 inch and the other is 2 inches, even though the absolute differences are the same. By the way, line 1 is 2.1 units, line 2 is 2.2, line 3 is 2.3, line 4 is 2.4, and line 5 is 2.5 units.

Fig. 3.5 Position along a Common Scale

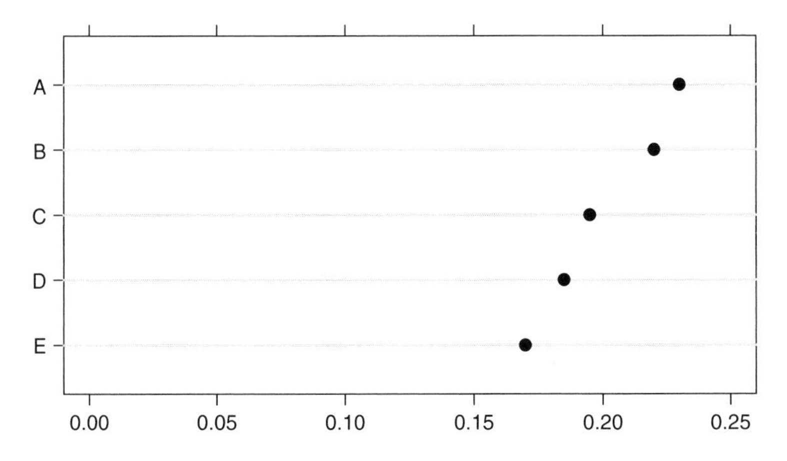

The dot plot shown in Figure 3.5 allows us to make judgments of positions along the common horizontal scale. Experiments by Cleveland and McGill (1984) have shown that this is the most accurate of the elementary graphical tasks. The dot plot was designed to take advantage of the knowledge gained from these experiments on perception and decoding information from graphs.

Fig. 3.6 Position along Identical, Nonaligned Scales

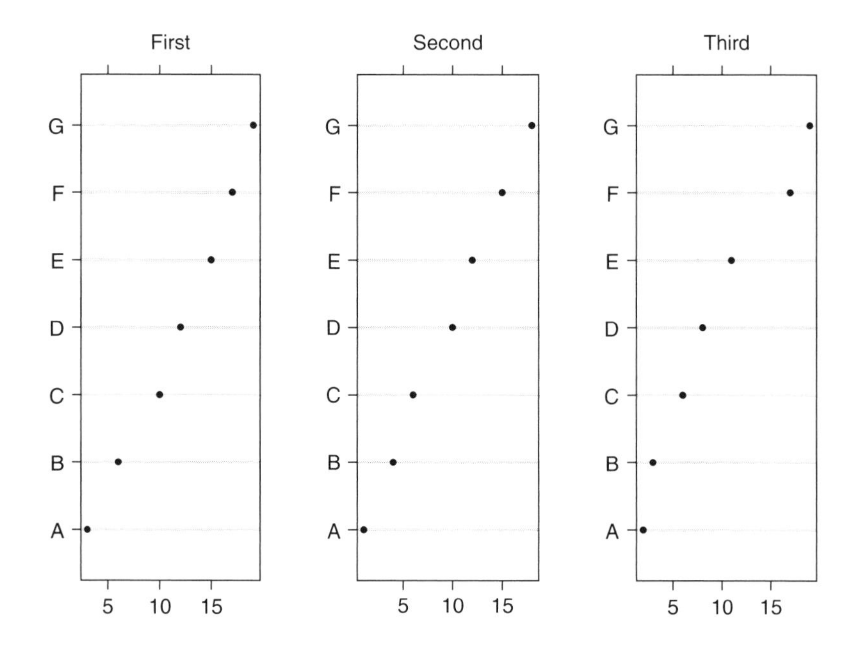

Note that the horizontal scales in Figure 3.6 are the same for the three cases shown. Within the same panel we judge position along a common scale. To compare values on separate panels, we compare positions along identical but nonaligned scales. We make these judgments very accurately.

Multipanel displays are extremely useful when we have more than two variables. Each panel shows two variables for one value of the third variable. For example, if the third variable is countries and we have data for England, France, and Italy, there would be one panel for each country. Multipanel displays are discussed in Chapter 5.

Fig. 3.7 Slope Judgments

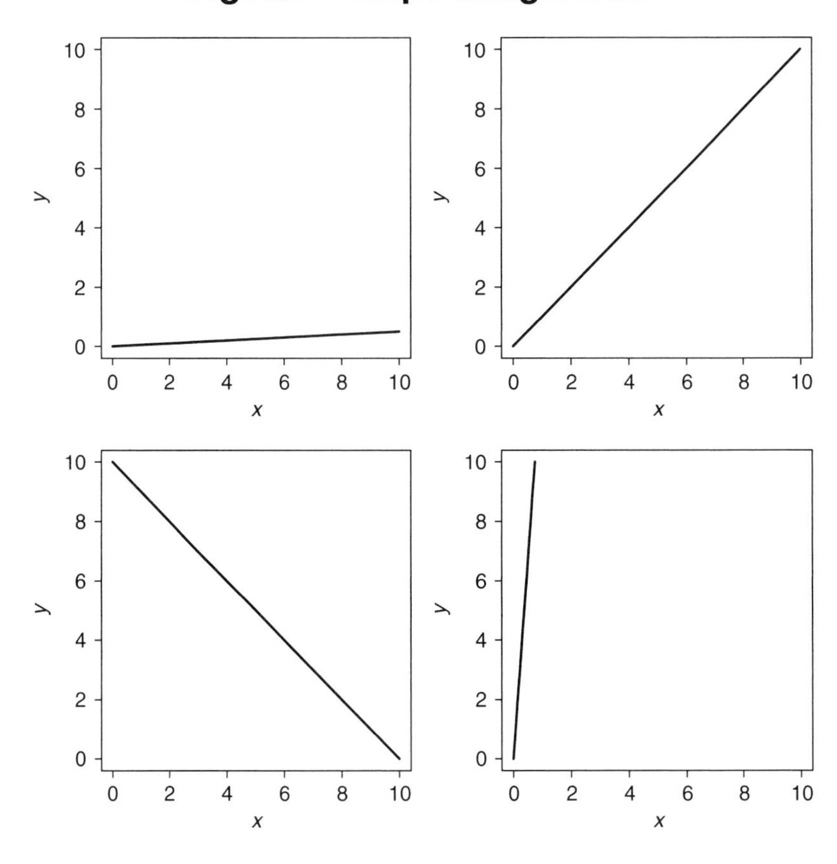

Readers make angle judgments to determine slopes, but we don't judge angles very accurately. The accuracy of judgments of slopes of line segments depends on the angle with the horizontal. Poor accuracy results from angles close to 90°. We judge angles near 45° most accurately. Included in Chapter 7 is a technique called *banking to* 45° that will tell you how to make use of this knowledge.

3.2 ORDERED ELEMENTARY TASKS

Page 47 shows the elementary tasks for decoding quantitative information in alphabetical order. The following list shows the same tasks in order of our ability to perform them accurately:

1. Position along a common scale
2. Position along identical, nonaligned scales
3. Length
4. Angle - slope
5. Area
6. Volume
7. Color hue - color saturation - density

in a range

Fig. 3.8 Detection

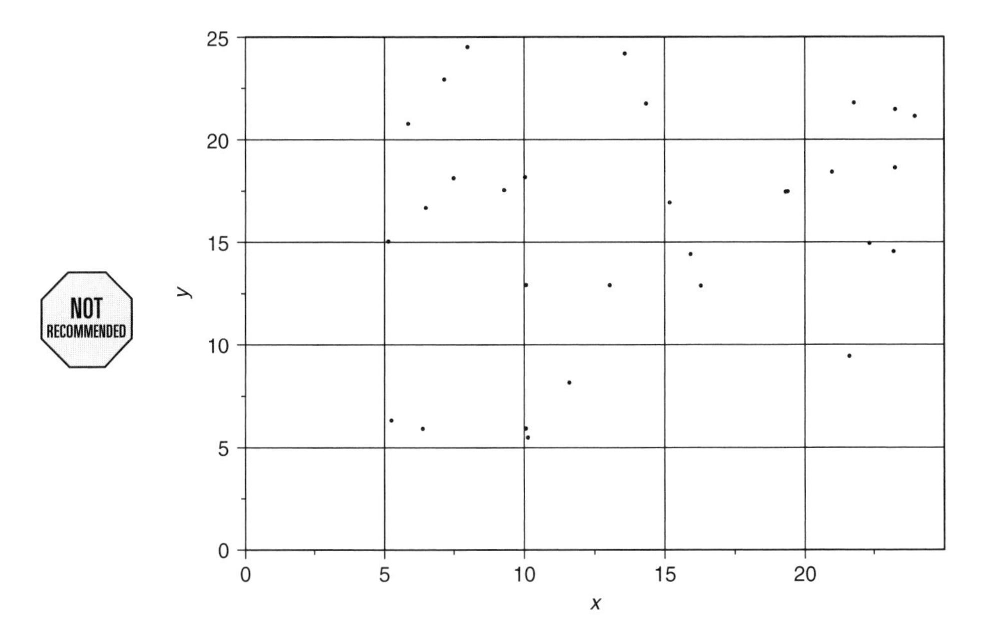

3.3 ROLE OF DISTANCE AND DETECTION

Distance and detection also play a role in our ability to decode information from graphs. The closer together objects are, the easier it is to judge attributes that compare them. As distance between objects increases, accuracy of judgment decreases. It is certainly easier to judge the difference in lengths of two bars if they are next to one another than if they are pages apart.

Before we can perform any of the elementary tasks, we must be able to detect the data. We often cannot do so if data points overlap one another; are hidden in the axes, tick marks, or grid lines; or are too light to see. Figure 3.8 illustrates some of these problems. Additional examples of data hidden by other graphical elements are provided in Chapter 6.

SUMMARY

Creating a more effective graph involves choosing a graphical construction in which the visual decoding uses tasks as high as possible on the ordered list of elementary graphical tasks while balancing this ordering with consideration of distance and detection.

4 Some More Effective Graphs in One or Two Dimensions

Chapter 2 showed how poorly chosen graphs can confuse and mislead. In this chapter and the next we illustrate the power of graphs by showing techniques that enable understanding of data that is not attainable in any other way. Some of these techniques have been available since the days of Playfair. Others are new. But all provide the basis for greater understanding of data. Understanding data better leads to better decisions, which can result in better policy or higher profits. Chapter 4 focuses on one or two variables, and Chapter 5 continues the discussion for three or more variables.

Creating More Effective Graphs, by Naomi B. Robbins
ISBN 0-471-27402-X Copyright © 2005 John Wiley & Sons, Inc.

We begin with graphs of one variable (*univariate*), which are used when there are only x values. These graphs show how the values are shaped, clustered, or spread. Some of these techniques are more effective than others, but none mislead when used properly. Next we present a comparison of univariate distributions. Graphs of two variables (*bivariate*) are used when we have x and y values. They show how x and y are related. Time series are a special type of bivariate graph used when measurements are taken over time.

Fig. 4.1 State Areas: Strip Plot

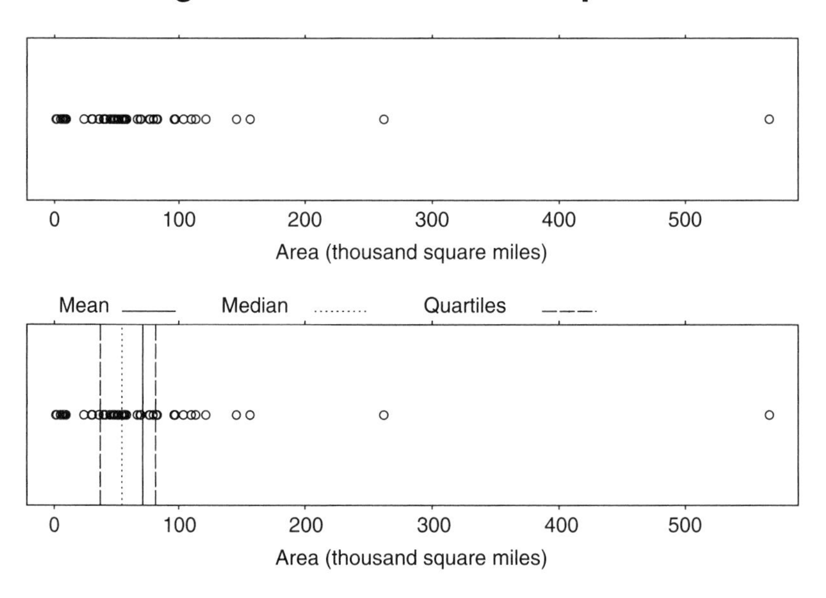

4.1 DISTRIBUTION OF ONE VARIABLE

4.1.1 Strip Plots

A *strip plot* shows the distribution of data points along a numerical axis; it is also called a *one-dimensional scatterplot, one-dimensional data distribution graph*, or *point graph*. The top strip plot in Figure 4.1 shows the distribution of the areas of the 50 states of the United States. It shows clearly the range of the areas and where most of the values lie, but not much more. The bottom strip plot includes some summary statistics; the mean is shown as a solid line, the median as a dotted line, and the 25th and 75th percentiles as dashed lines. Strip plots are sometimes used in the margins of two-dimensional displays to show the distribution of each variable separately.

Fig. 4.2 State Areas: Dot Plot

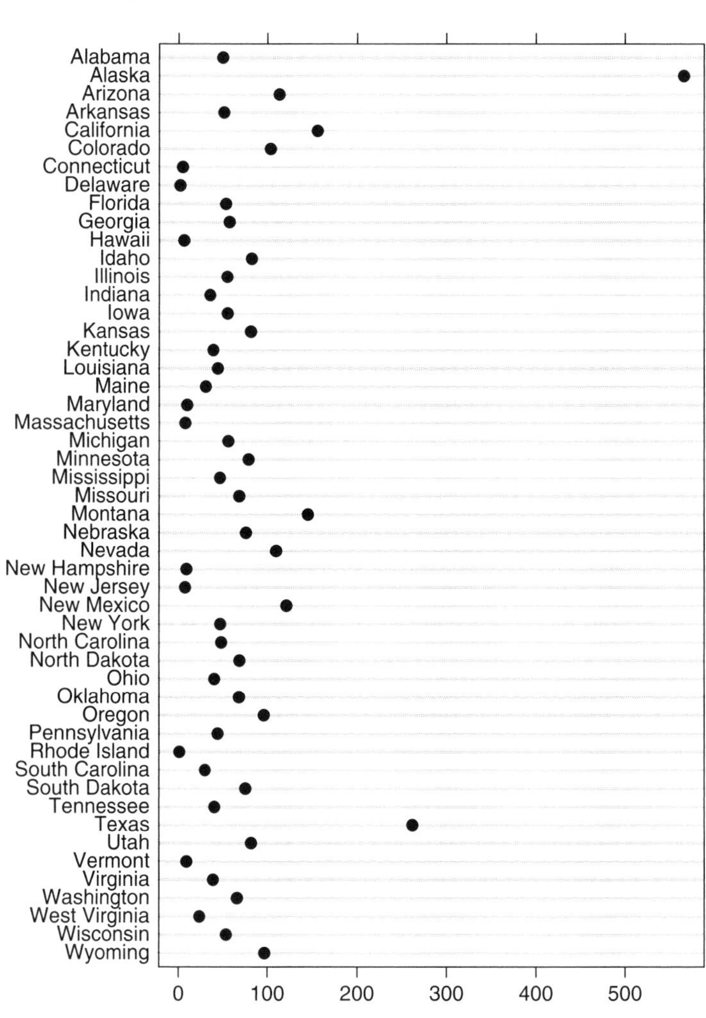

4.1.2 Dot Plots

Figure 4.2 shows the areas of the 50 U.S. states plotted using a dot plot. Dot plots were introduced by Cleveland (1984) after extensive experimentation on human perception and our ability to decode graphical information. Since the judgments the reader makes when decoding the information are based on position along the common horizontal scale, these plots display data effectively. Since it would be very difficult to fit the names of the states on the horizontal axis, dot plots place them on the vertical axis, and the quantitative variable, area in thousands of square miles, on the horizontal axis. Notice how much more informative the dot plot is than the strip plot. Although the data appear clearly, we can still improve this chart.

Fig. 4.3 State Areas: Ordered by Size

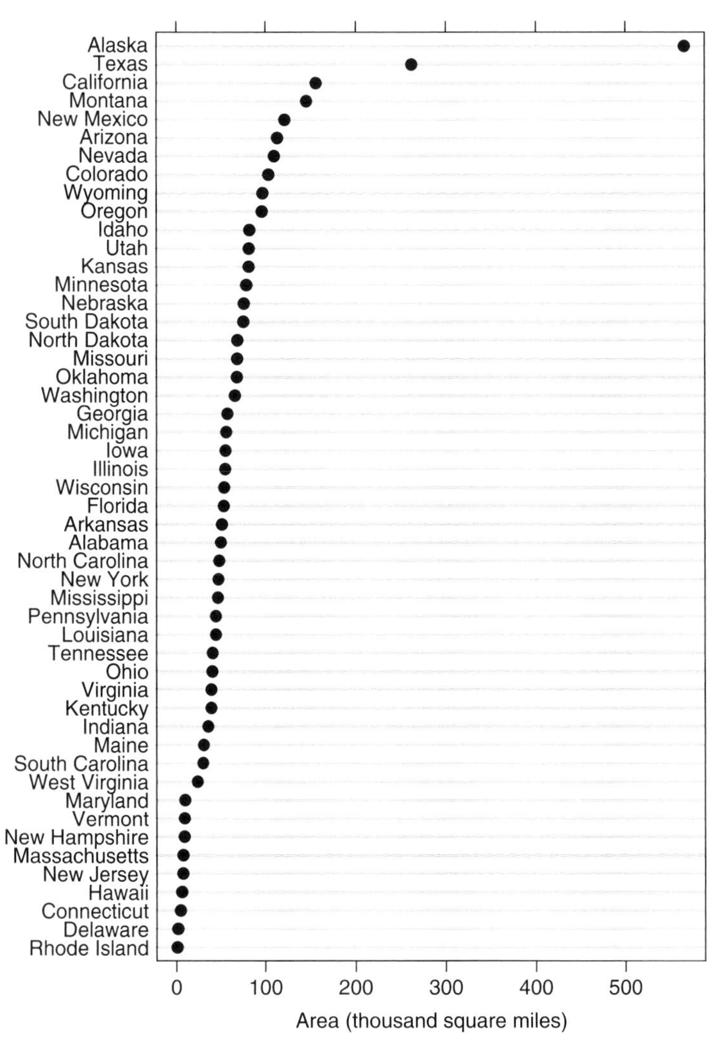

The states appear alphabetically in Figure 4.2. In Figure 4.3 they are listed in order of size. This presentation is much more informative. It is much easier to answer such questions as: "How many states are smaller than Indiana?" or "What is the median size of a state?" This figure could be improved even more. Notice that Texas and Alaska are so much bigger than the other states that most of the data are on the left side of the chart. Figure 4.4 shows how to handle this problem, but it is not the best choice for every situation. As in any form of communication, we must know our audience and tailor what we say to be appropriate for that audience, the readers of the chart.

Fig. 4.4 State Areas: Logarithm with Base 2

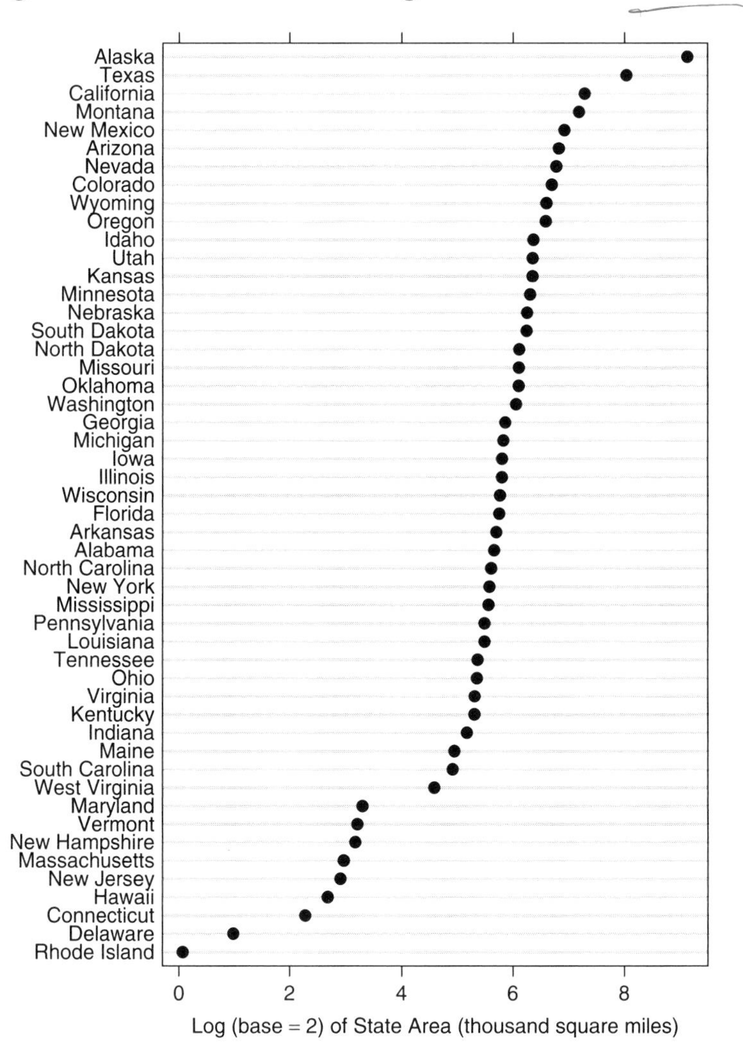

A logarithmic scale makes it possible to plot values with too wide a range for a linear scale. You have probably seen graphs plotted on a logarithmic scale, with the axis labeled 10, 100, 1000, and so on, in financial reports. Let's review what we mean by *logarithms*. If $10^2 = 100$, then $\log_{10}(100) = 2$. $y = \log_b x$ means that b is raised to the exponent y in order to get x.

Bases other than 10 are also useful. Ten is useful when the data range over several orders of magnitude. A base of 2 is useful for plots when we want to spread the data over a smaller range. Figure 4.4 shows the state areas on a logarithmic scale with a base of 2. This allows us to see details that were not as clear on the linear scale: for example, the large jump in size from Maryland to West Virginia.

Fig. 4.5 State Areas: Top Axis Labeled with Original Scale

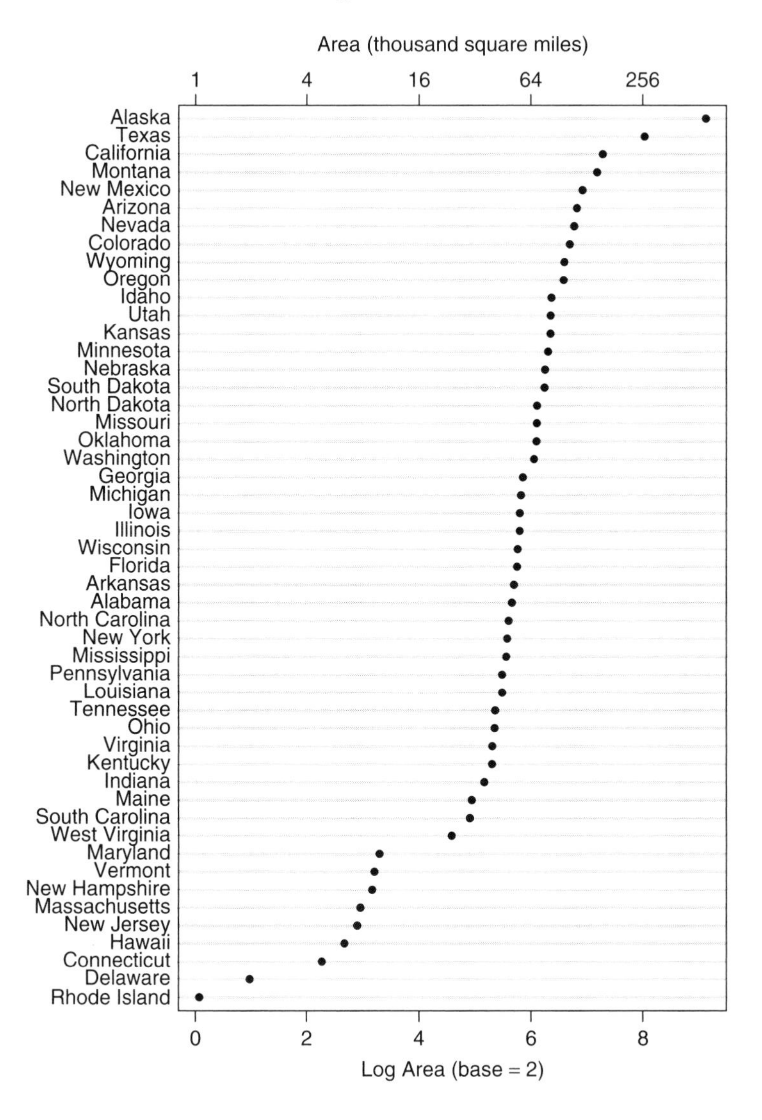

Logarithmic scales are useful for understanding multiplicative factors. From Figure 4.5 we see that the state of Washington is approximately $2^6 \times 1000$, or 64,000 square miles, whereas Texas is approximately $2^8 \times 1000$, or 256,000 square miles. That tells us that Texas is the size of Washington times 2^2, or about four times the size of Washington, since $2^8 = 2^6 \times 2^2$.

The horizontal axis is labeled with the logarithmic scale. It is useful also to label the figure with the original scale, to make it easier to understand. We do that on the top axis of Figure 4.5. Note the connection between the labels on the bottom and top; 2 raised to the value of the bottom label gives the value of the top label. More information on logarithmic scales is given in Chapter 7.

Fig. 4.6 Families Exhibition: Histogram

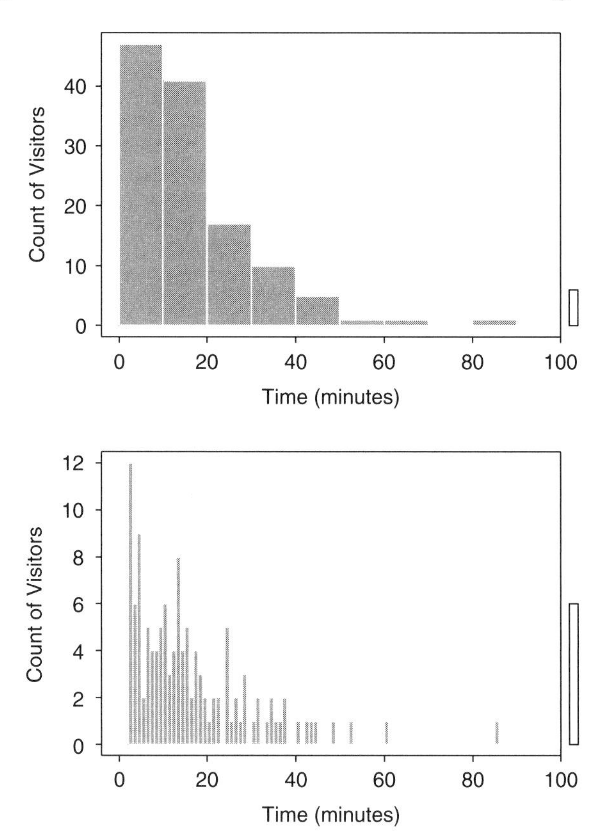

4.1.3 Histograms

Histograms show the distribution of a set of data. Serrell (1998) examined the number of minutes that visitors spent at museum exhibitions. To draw a histogram, the data are grouped into bins or intervals. For example, in the top chart in Figure 4.6, which shows the time that visitors spent at an exhibition named "Families," all times up to and including 10 minutes are in the first bin, times from 10 to 20 minutes are in the second bin, and so on. Then the count of the number in each bin or the percent of the total in each bin is plotted. There is a trade-off between showing detail or showing a better overall picture. The top figure shows the shape of the distribution more clearly, and the bottom figure shows more detail. Histograms do a reasonable job of showing the shape of one data set but are not very useful for comparing distributions.

Note that the scales of the two histograms are not the same. Since there are more bins in the bottom figure, there are fewer visitors in each bin. To help visualize this difference in scales, the rectangles on the right both have a height of six visitors.

Fig. 4.7 Avoid Misleading Histograms

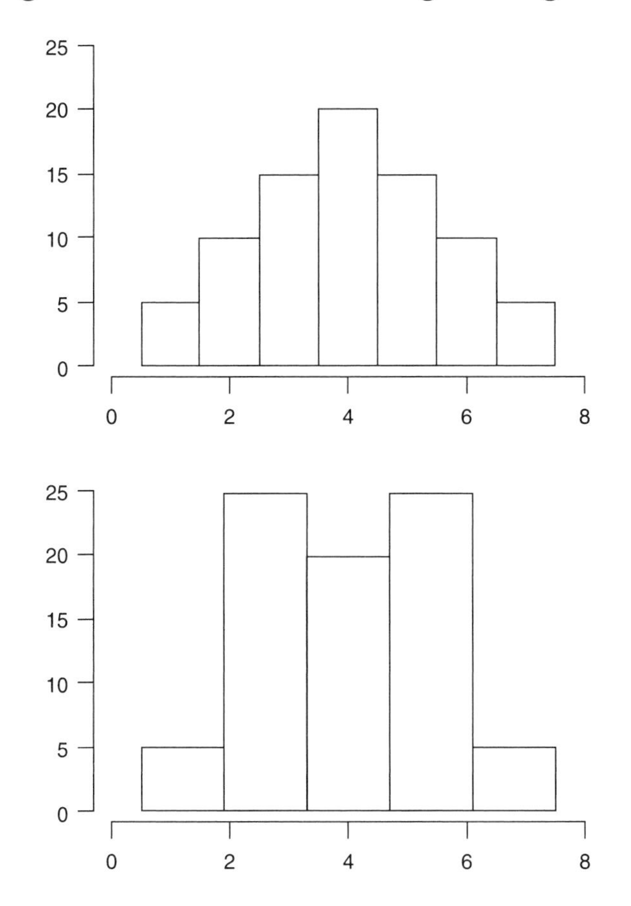

Some software packages for drawing histograms allow the user to make an unreasonable choice of the number of bars to use. For example, suppose that there are seven possible data values and that all are integers. The value "1" appears five times, "2" appears 10 times, and the others appear as shown on the top chart in Figure 4.7.

The x axis of the top chart goes from 0.5 to 7.5 with a range of 7, so that each bar contains exactly one of the integers. However, as the bottom chart shows, if the user requests five bars, an unreasonable number for these data, each bar has a width of 1.4, which is 7 divided by 5. The first bar goes from 0.5 to 1.9, including the five occurrences of 1. The second bar goes from 1.9 to 3.3, including the 10 occurrences of 2 and 15 occurrences of 3. This creates a very misleading impression of the shape of the distribution, as shown in the bottom chart.

Fig. 4.8 Judith Leyster Exhibition: Strip Plot

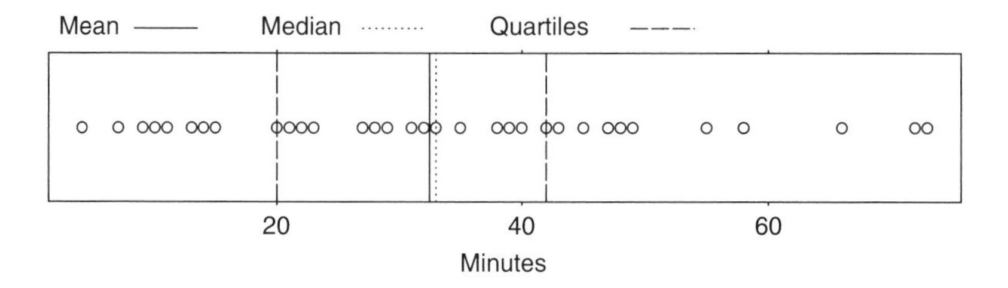

4.1.4 Jittering

The strip plot in Figure 4.8 shows the number of minutes visitors spent at an exhibition of the artist Judith Leyster. The data were collected to the nearest minute. There are 49 observations:

4	7	7	9	10	10	11
11	13	14	15	15	20	21
22	22	23	27	27	28	28
29	31	32	33	33	35	38
38	39	40	40	40	40	42
42	42	43	45	47	48	48
49	49	55	58	66	72	73

Notice that there are many repeat values, so that some plotting symbols overlap one another.

Fig. 4.9 Judith Leyster: Jittered Strip Plot

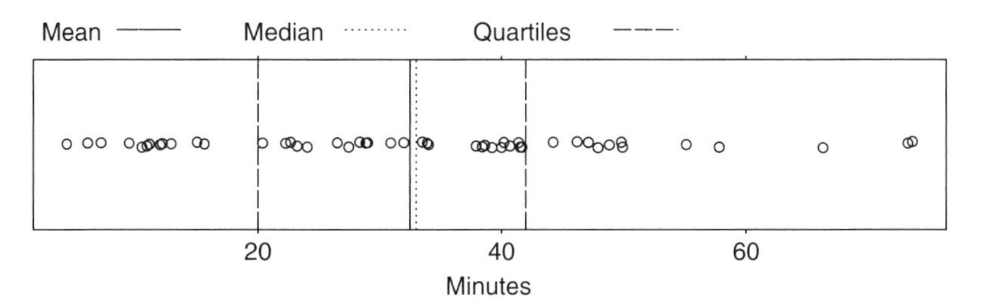

To make the data points distinguishable, we have added random noise to the data before plotting Figure 4.9. This technique, called *jittering*, moves the data points a small, random amount from their original positions so that they no longer overlap. Many software packages allow you to jitter your data. If yours does not, you can generate random numbers with a small variance or spread to achieve this effect.

There are a number of other solutions to this problem of overlapping data points, which appear on page 165 and in Cleveland (1994).

Fig. 4.10 Museum Exhibitions: Jittering

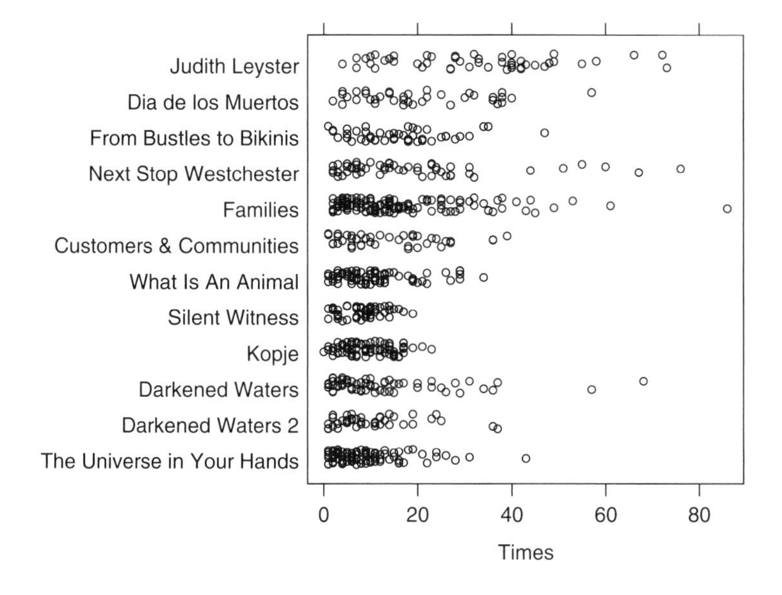

Figure 4.10 shows a strip plot of the time that visitors spent at 12 exhibitions, with the data jittered. It gives a much better indication of the distributions than would a plot without jittering.

Strip plots are a reasonable choice if the number of observations is not too large or if the observations are spread out. Even with jittering, the plots will become indecipherable for a large number of observations with a small range of values. In such a case, histograms are preferable as long as only one distribution is to be plotted. Box plots, such as Figure 4.11, are better for comparing distributions.

4.2 COMPARING DISTRIBUTIONS: BOX PLOTS

It is often useful to compare two or more sets of numbers. We might, for example, compare mean SAT scores of males and females. The mean alone, however, provides limited information since very different distributions or spreads of data may have the same mean. Let's suppose that John Speaker gives lectures at conferences. The conference collects feedback from attendees on a scale of 1 to 5 (with 5 being excellent). John is told that his mean score was 3 and that 100 people attended his talk. He was also told that the mean score for the entire conference was 4. One possible scenario is that all 100 attendees gave John a score of 3. That tells John that he should work on improving the contents or delivery of his presentation. Another possible scenario is that 50 people gave him a score of 5 and the other 50 gave him a score of 1. If 50 attendees gave a score of 5, it probably is already a very good presentation. However, many people attended who might not be interested in the subject or expected something different. In this case, John or the conference committee should work on improving the description and marketing materials for his session so that it attracts the right audience. Just supplying the mean gives John no clue as to which of these scenarios was actually the case.

Graphs allow us to see an entire set of data. The strip plot of visitors' times at museum exhibitions shown in Figure 4.10 enables us to compare visitors' behavior at different exhibitions. We see that the times at Judith Leyster are more spread out than at some of the others, which have darkened areas indicating a large number of visitors in a small time interval. Some of the exhibitions had visitors who stayed much longer than most; Silent Witness and Kopje did not. Box plots are excellent for comparing distributions.

Fig. 4.11 Museum Exhibitions: Box Plot

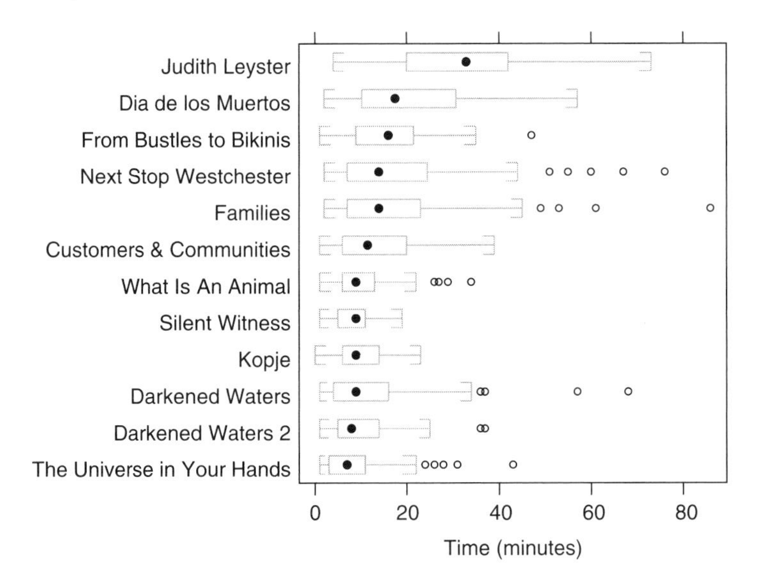

Box plots are much more effective than histograms or strip plots for comparing distributions of more than one data set. Figure 4.11 shows the number of minutes that visitors spent at 12 exhibitions. The filled circles represent the median time for each exhibition, the rectangles show the middle half of the data,[1] and the open circles represent outliers. The whiskers at the ends of the lines represent the range of the data without outliers. The range and outliers are defined more precisely on page 91.

John Tukey (1977) introduced box plots in the 1970s. You are giving away your age if you confess that you haven't seen them before, because today, box plots are included in math classes in many middle and high schools.

[1] Middle half: the data from the 25th percentile to the 75th percentile.

Fig. 4.12 Description of Box Plot

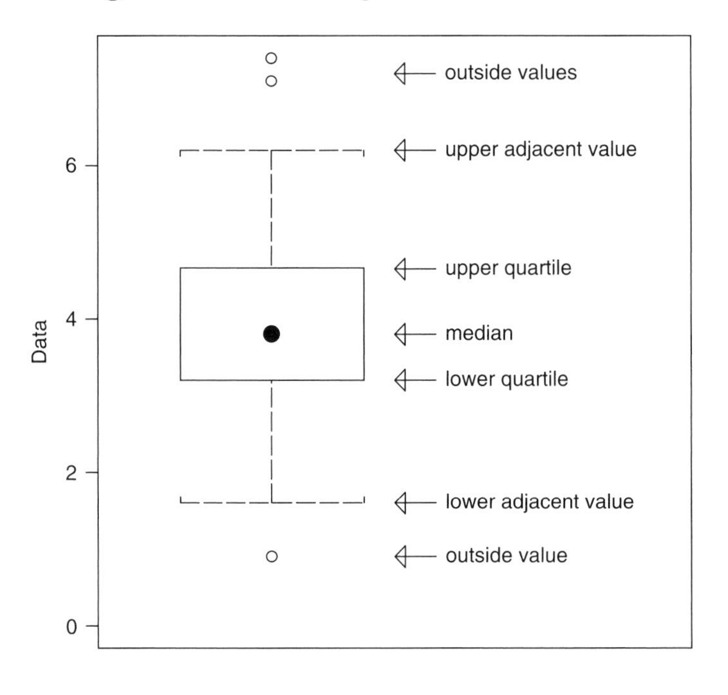

Box plots are often called *box and whisker plots*. The middle circle is the median. The rectangle shows the *interquartile range* (IQR); it goes from the first quartile (the 25th percentile) to the third quartile (the 75th percentile). The whiskers go from the minimum value to the maximum value unless the distance from the minimum value to the first quartile is more than 1.5 times the IQR. In that case the whisker extends out to the smallest value within 1.5 times IQR from the first quartile. A similar rule is used for values larger than 1.5 times IQR from the third quartile. A special symbol shows the values, called *outliers*, which are smaller or larger than the whiskers. Open circles are used here. Other styles, such as line segments for the medians, are also used.

Fig. 4.13 Museum Exhibitions: Styles of Box Plots

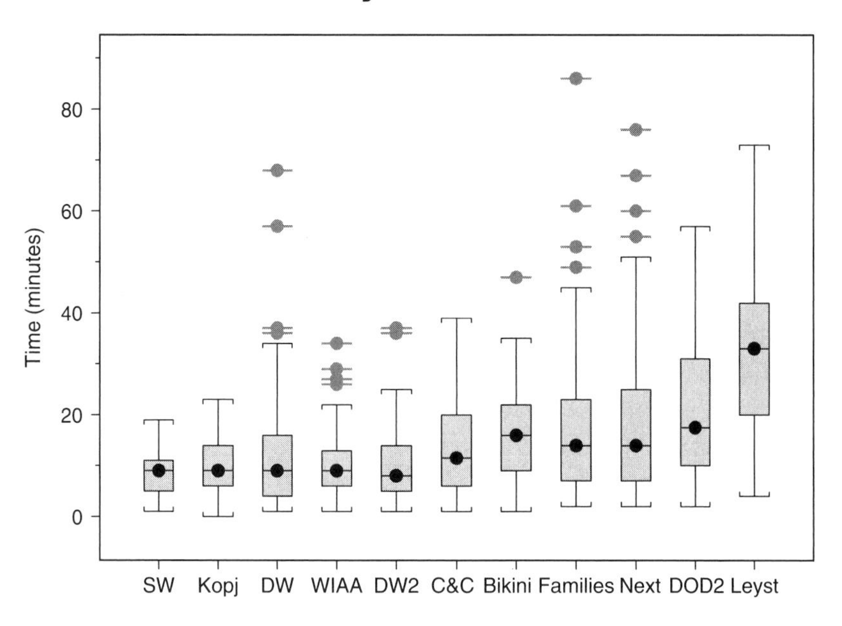

There are a number of style choices for box plots. In Figure 4.13 we show the museum exhibition data in a vertical box plot. The problem is that there is no room for the names of the exhibitions, so we have to abbreviate them. In addition to the vertical orientation, other style changes include horizontal lines with the symbols for the median and outliers as well as shaded boxes.

Some software packages may use different definitions for outliers than the one used here. The definitions here are consistent with those of Cleveland (1994) and the S language (Becker, Chambers, and Wilks, 1988). Check the documentation of the software you use.

Fig. 4.14 *Paying Attention* Data: Scatterplot

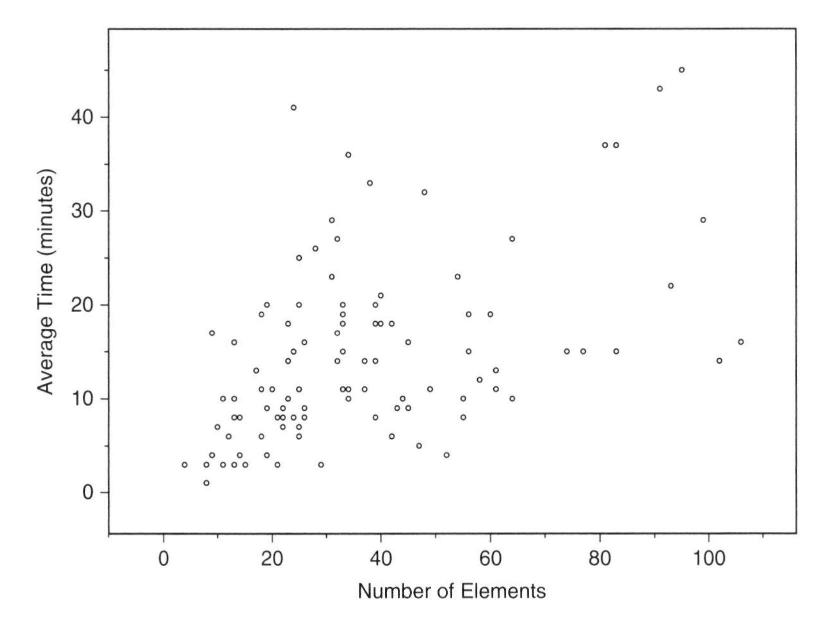

4.3 RELATIONSHIP OF TWO VARIABLES: SCATTERPLOTS

A *scatterplot* is a basic graph form that is used for two variables that are both quantitative. It has many names: *scattergram*, *xy plot* (Excel), and even *dot chart* (Zelazny, 1996). A well-drawn scatterplot is a useful, tried-and-true method for seeing the relationship between two variables, the distribution of points in the plane, clusters of points, and outliers.

Serrell (1998) examined characteristics of museum exhibitions to develop a methodology for evaluating the effectiveness of exhibitions. Figure 4.14 shows the average time and the number of elements for 103 exhibitions. *Elements* refer to activities or planned stops at an exhibition, and *time* refers to the amount of time that a visitor spent at the exhibition. It is clear that the average time increases with the number of elements, but we might be interested in knowing if this increase is linear. A linear fit provides a simple model for predicting y (here, average time) given x (here, the number of elements). Principles for choosing scales, the number of tick marks and labels, grid lines, and other graphical elements are given in Chapters 6 and 7.

Fig. 4.15 *Paying Attention* Data: Regression Line

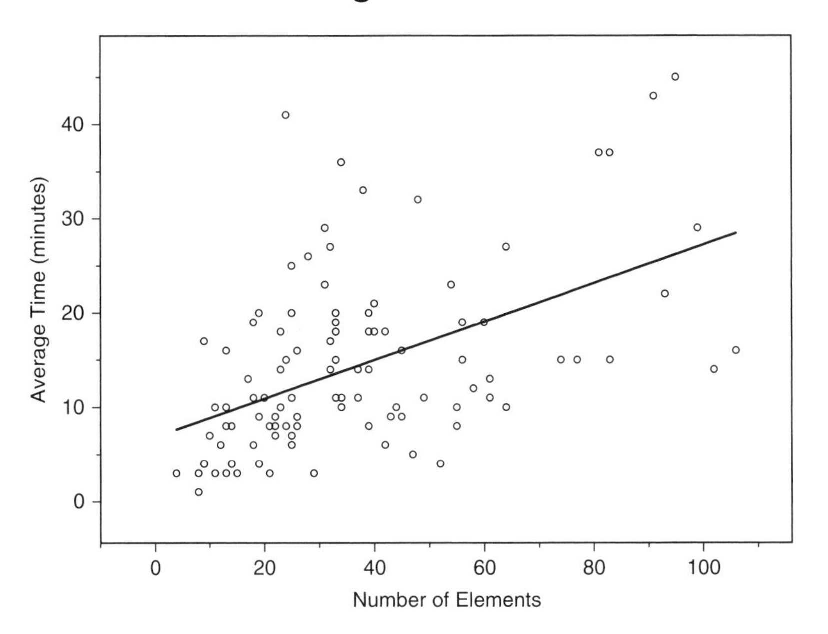

To see if the relationship between the number of elements and the average time is linear, a regression line is fit to the data using the *method of least squares*.[2] For any x value, the vertical distance between the data and the line is called the *residual*. The *least squares line* is the line that minimizes the sum of squares of these residuals. The reader is referred to Cleveland (1993) or any introductory statistics text to learn more about regression analysis and how to determine if a linear fit is a good fit.

[2] Readers with no background in statistics may want to skip our discussion of Figures 4.15 and 4.16. When you see the term *loess* in Chapter 7, it refers to a procedure that statisticians use to fit a curve to data or to find trend lines.

Fig. 4.16 *Paying Attention* Data: Loess Fit

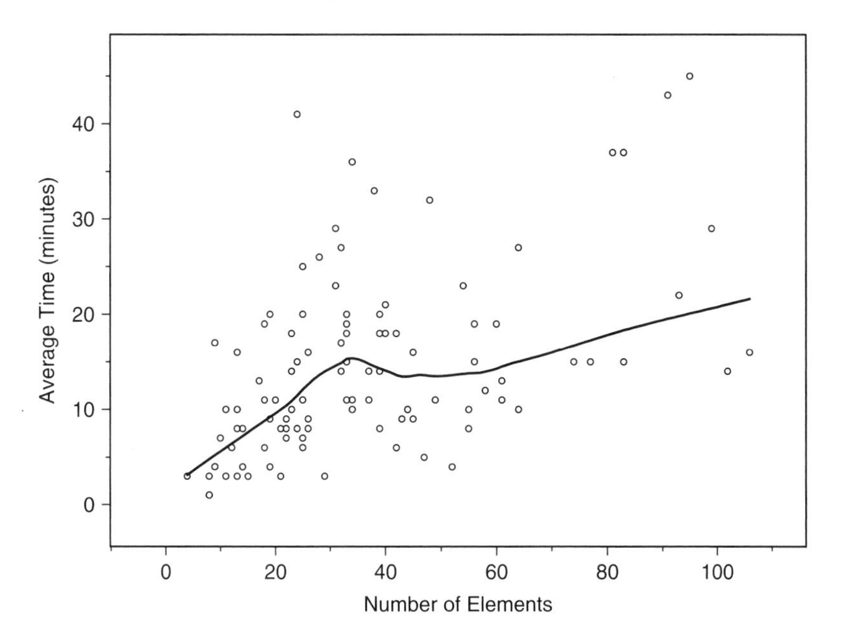

When we fit a regression line, we assume that the data are linear and then fit the best line. But what if we aren't sure if the fit is linear? There is a class of procedures called *scatterplot smoothers* that fit a curve to data points locally. That means that the curve fit at any point depends on that point and other data in its neighborhood. The curve fit[3] in Figure 4.16 uses *loess*, which is a technique for locally weighted regression smoothing. The curve at point x is found by fitting weighted linear or quadratic functions to the data points near x. A steeper slope indicates a higher rate of increase. We see that although average time increases with the number of elements, the rate of increase is higher for lower numbers of elements.

[3] Both this curve and the regression curve describe this set of exhibitions; they should not be generalized because the exhibitions were not chosen randomly.

Fig. 4.17 Visitors at St. Louis Science Center

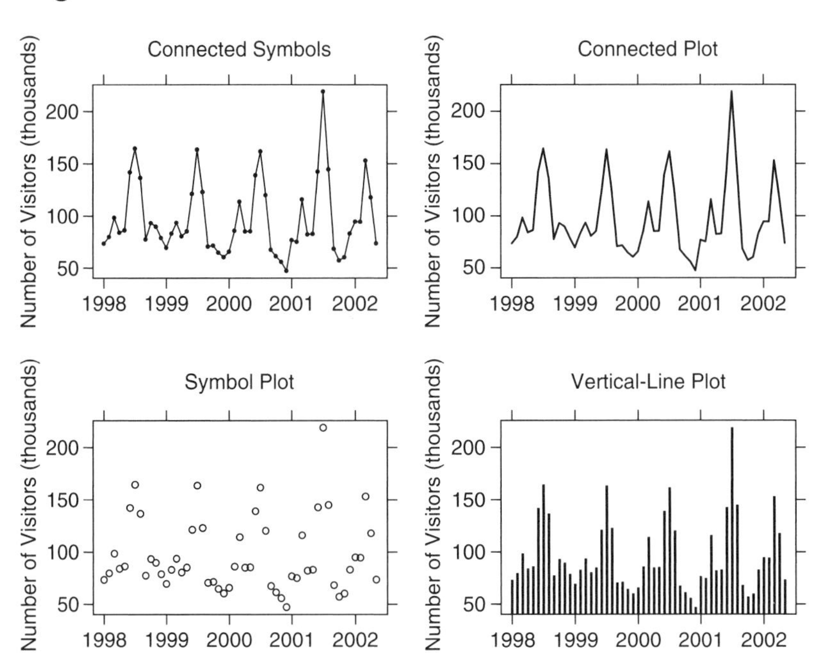

4.4 TIME SERIES

Time series result from measurements taken over time. Graphs of time series are among the most frequently occurring graphs. The four plots in Figure 4.17 show the same data, the number of visitors at the St. Louis Science Center (SLSC) from January 1998 through May 2002 (Tisdal, 2002), in different forms. The top left uses connected symbols, allowing us to see both the individual data points and their ordering over time. The bottom left displays symbols only, which is useful if we wish to fit a trend to the data. The connected plot in the top right is appropriate when it is not important to see the individual data points. Finally, the vertical-line plot enables us to pack the series tightly along the horizontal axis and is used when we need to see individual values and short-term fluctuations. A connected plot with or without symbols is often called a *line plot*.

Fig. 4.18 SLSC Visitors: Month Plot

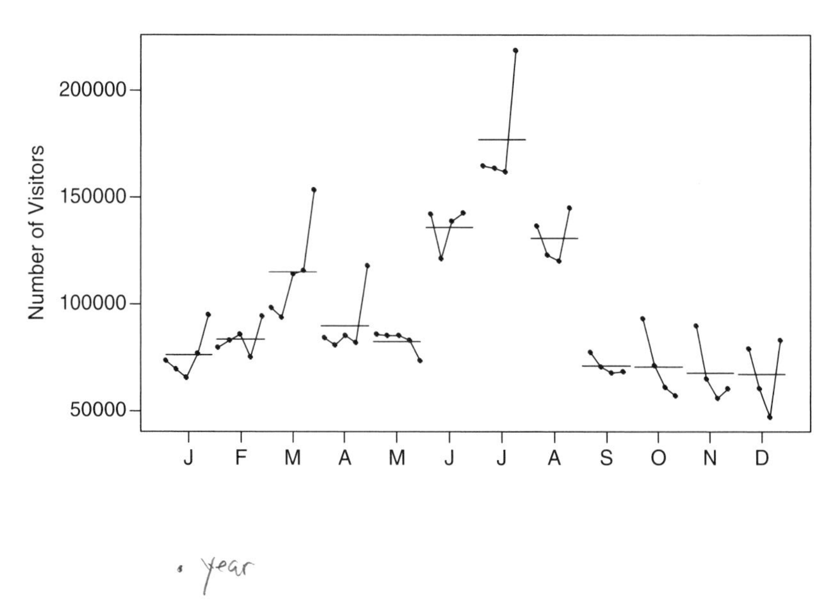

A casual look at the attendance data time-series plots does not indicate any strong trend but does show a strong cyclical pattern. *Month plots* (Figure 4.18) allow us to see the behavior of subseries. We first plot all the January values, then the February values, and so on. The horizontal lines represent the means for each month. They indicate the overall monthly pattern of the data, while the subseries plots give the behavior by year for a given month. Thus we discover characteristics of the data that are difficult to see in the full series: for example, that attendance has decreased year after year in May and the fall months. Plotting all the subseries on the same graph lets us see how changes in a subseries compare with the overall pattern of the data. Month plots are also called *cycle plots*.

Fig. 4.19 Carbon Dioxide Data

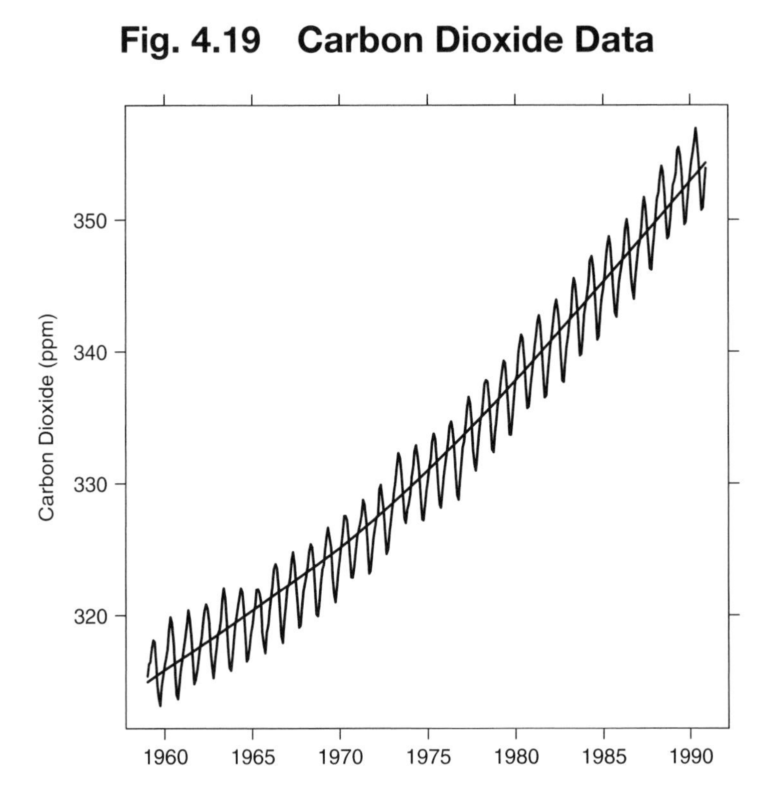

The data in Figure 4.19 show the monthly average concentrations of carbon dioxide in the atmosphere as measured at the Mauna Loa Observatory in Hawaii. Again, there is a strong seasonal effect, but in this case there is also an increasing trend. The trend shows that the carbon dioxide levels have been increasing over time. The seasonal cycles result from the changes in foliage in the northern hemisphere. The carbon dioxide in the atmosphere increases when the foliage decreases in the fall, and decreases when the leaves bloom again in the spring. The loess curve in this example shows that not only is the amount of carbon dioxide increasing, but the rate of increase is also increasing, commonly referred to as *accelerating*. To see this, hold a straightedge along the bottom part of the curve. You will see that the top part of the curve bends up from your straightedge, indicating an increasing rate of carbon dioxide in the atmosphere.

Fig. 4.20 Carbon Dioxide Data: Month Plot

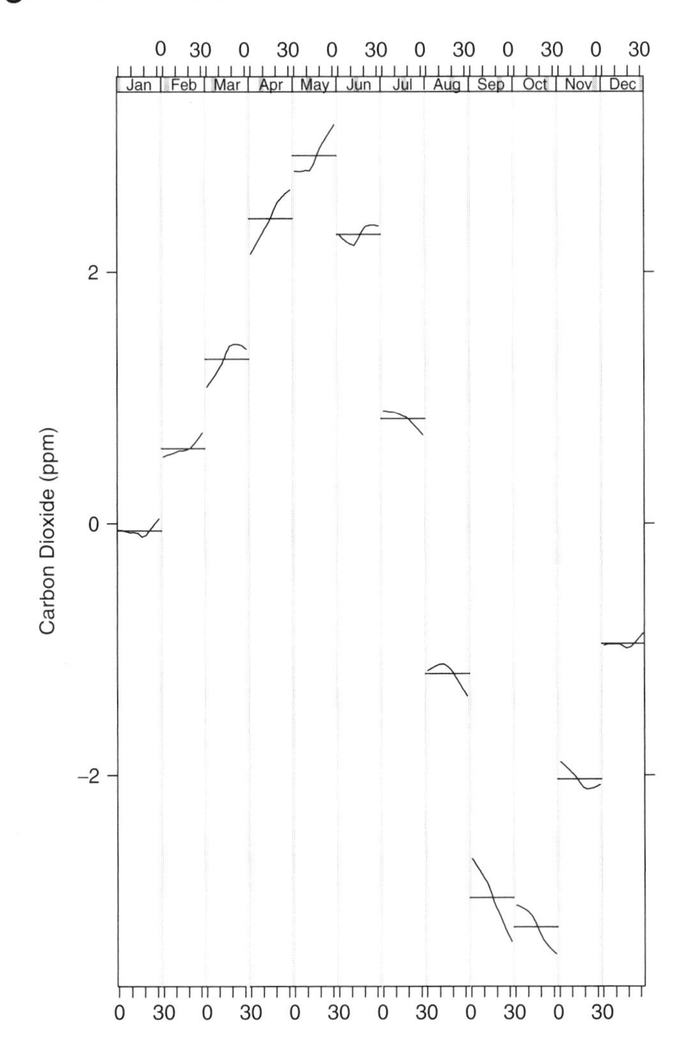

30 years for each month

Figure 4.20 shows an alternative presentation of a month plot; it displays the seasonal component of the carbon dioxide data (Cleveland and Terpenning, 1982). This shows clearly that the annual cycle reaches a maximum level in May and a minimum level in October. It also clearly shows that the levels of carbon dioxide in the spring are increasing, whereas those in the fall are decreasing, so that the range of the cycles is increasing. The data set includes measurements of carbon dioxide for over 30 years (from 1959 through 1990).

Fig. 4.21 Used Car Price: Line Graph

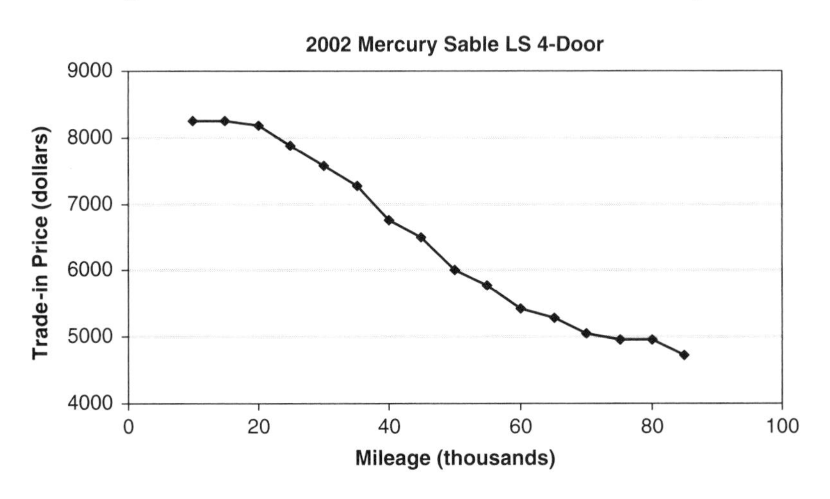

Fig. 4.22 Mountain Height Data: Line Graph

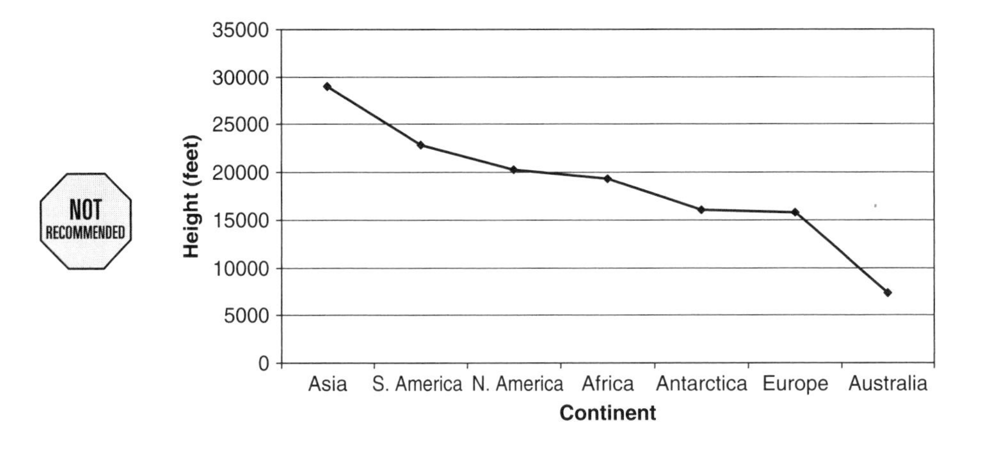

4.5 LINE GRAPHS

The common name for a connected plot or connected symbol plot, introduced in Section 4.4, is *line graph*. Other names are *fever graphs* and *thermometer graphs*. Line graphs occur most frequently with time series; however, they may be used with other quantitative variables on the x axis, such as temperature, age, or mileage. Figure 4.21 gives the trade-in price as of early 2004 of 2002 Mercury Sables with varying mileages. The line helps to interpolate values if you are interested in a mileage for which no point is plotted.

It is common to see line graphs for nonordered variables as well. I do not recommend this practice. The heights of the highest mountain in each continent (Sutcliffe, 1996) are shown in Figure 4.22. In this case, interpolation does not make sense. What does it mean to find a value halfway between apples and oranges, or, as in this case, Asia and South America? Although the two figures look quite similar, the horizontal axis displays a quantitative variable in Figure 4.21 and a categorical variable in Figure 4.22.

Fig. 4.23 Mountain Height Data by Mountain

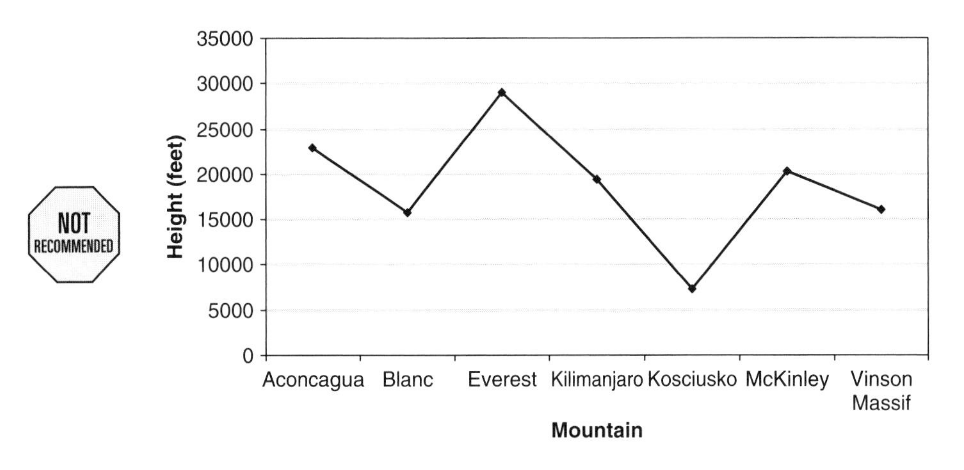

Fig. 4.24 Mountain Height Data by Continent

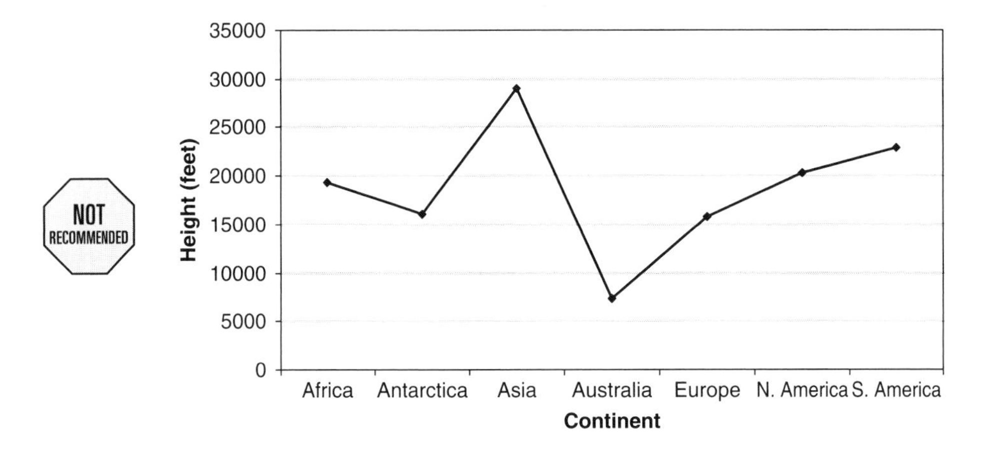

Another reason that I don't like line graphs for data that do not have a natural ordering is that the visual impression we get depends so heavily on the ordering of the categories. The same data are plotted in both Figures 4.23 and 4.24, but in Figure 4.23 the label gives the name of the mountain and in Figure 4.24 the label gives the name of the continent. Both are ordered alphabetically. It is difficult to accept the fact that these are the same data. I prefer dot plots for these data since you connect the value to its label rather than to another value.

Some argue that the lines help the eye to follow the points, especially when there are multiple data sets superposed on one figure that need to be grouped visually. However, in view of the disadvantages mentioned, I do not recommend line charts for unordered categorical data. An example of ordered categorical data is: strongly disagree, disagree, neutral, agree, and strongly agree. Here I do not object to line charts since interpolation might make sense and no one would reorder the categories in an arbitrary way.

An exception to my position occurs when many graphs are produced using the same categories. For example, several medical and/or psychological tests might be conducted on a large number of subjects with the results plotted in graphical form. If the categories are in the same order in all the graphs, connecting the lines might help the examiners look for patterns that are typical of the problems they are trying to identify.

4.6 COMMENTS

Some readers will complain that their software cannot produce all of the graph forms shown here. If you are one of them, you might be surprised: Your software may be more powerful than you realize. For example, Chapter 9 explains how to draw dot plots with Microsoft Excel. Even if your software cannot be used to draw some of the graph forms recommended here, many of the ideas can be adapted. For example, we showed how dot plots are often more effective than pie charts. Bar charts also are judged by position along a common scale and are preferable to pie charts for some small data sets. Therefore, two-dimensional bar charts may be used for some of these situations. There are a number of reasons why I prefer dot plots to bar charts. Bar charts get cluttered quickly and cannot handle as large a data set as dot plots can. We visualize areas as well as position with bar charts, whereas we visualize only position with dot plots. The importance of this will become apparent in Chapter 7 when we discuss whether zero must be included in scales.

To produce a graph similar to a month plot, you can create subseries manually and plot each one separately. Chapter 9 covers software freely downloadable from the Web that can produce all the graphs shown here.

SUMMARY

Dot plots, box plots, scatterplots, month plots, and the other graph forms described in this chapter are more effective than the everyday charts in Chapter 2 for describing or comparing data sets, seeing the relationship between variables, and observing trends and patterns over time.

5 Trellis Graphics and Other Ways to Display More than Two Variables

Fig. 5.1 Energy Data: Stacked Bar Chart

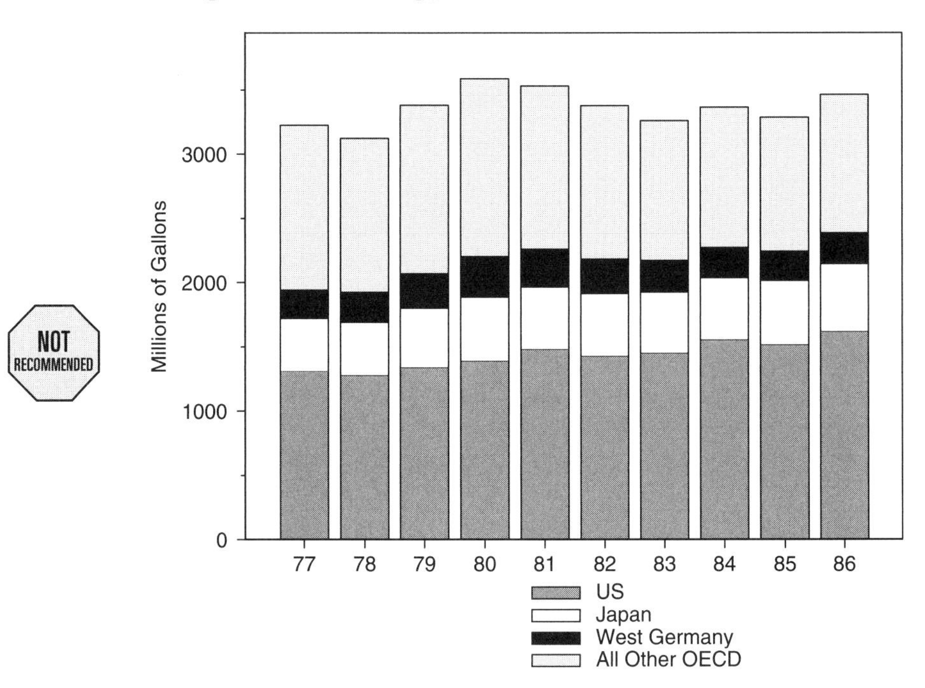

Creating More Effective Graphs, by Naomi B. Robbins
ISBN 0-471-27402-X Copyright © 2005 John Wiley & Sons, Inc.

A continuing problem in the display of quantitative information is the presentation of multivariate data on two-dimensional paper. Over the years, suggestions have been made that have often been more novel than effective. One recent and useful innovation is trellis display. In this chapter we present a glimpse of trellis display and other methods for displaying more than two variables.

5.1 ALTERNATIVE PRESENTATIONS OF THREE VARIABLES

5.1.1 Stacked Bar Chart

In the stacked bar chart in Figure 5.1 (repeated from Figure 2.11), the values for the United States are easy to see because we are judging position along a common scale. The values for the total are also easy to see. It is much more difficult to see trends in the other countries. There are three variables: year, gallons of petroleum, and country. The year and the gallons are quantitative, whereas the country is categorical. The next few pages include clearer alternative presentations of these data.

Fig. 5.2 Energy Data: Labeled Scatterplot

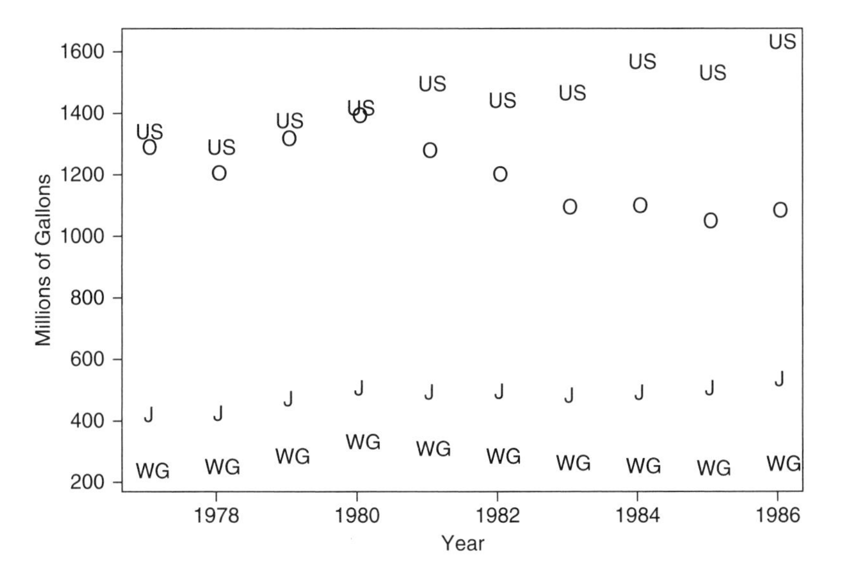

5.1.2 Labeled Scatterplot

The four countries in Figure 5.1 are superposed on the *grouped* or *labeled scatterplot* shown in Figure 5.2. Each country is plotted with a different symbol. The choice of symbols is discussed in Section 6.2.1. In this case I used letters suggestive of the countries since there is little overlap: US for the United States, O for all other OECD, J for Japan, and WG for West Germany. Figure 5.2 shows the trends better than the stacked bar chart (Figure 5.1) shows them. It is also easier to estimate the number of gallons for each country than with the stacked bar chart. If we connect the points, we have a multiple line chart. The four sets of data are clearly separated here, but that is not always the case.

Fig. 5.3 Energy Data: Trellis Display

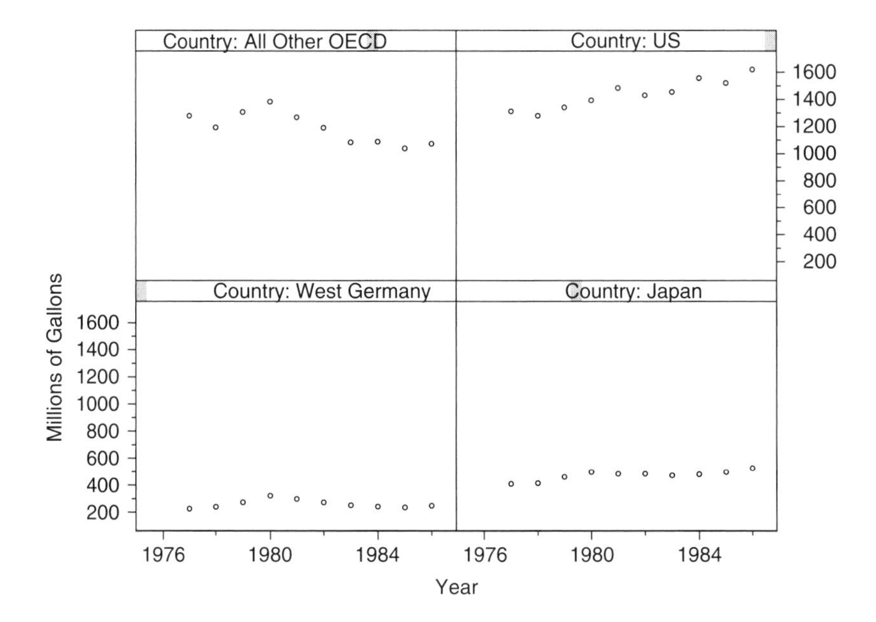

5.1.3 Trellis Display

The countries category has four values: US, Japan, West Germany, and all other OECD. Let's look at one country at a time and plot petroleum reserves versus years for that country. We call this *conditioning on the country* (e.g., the top right plot in Figure 5.3 is conditional on the country being the United States). The panels are ordered by their median levels of petroleum from bottom left to bottom right to top left to top right. The darker shading on the strip labels indicates this ordering; West Germany has the fewest gallons of petroleum and the United States the most. To make comparisons easier, each panel is plotted using the same scale. A panel can easily be added for the total if that is of interest. The display is termed *trellis* because the layout of panels is reminiscent of a garden trellis.

Fig. 5.4 Barley Data

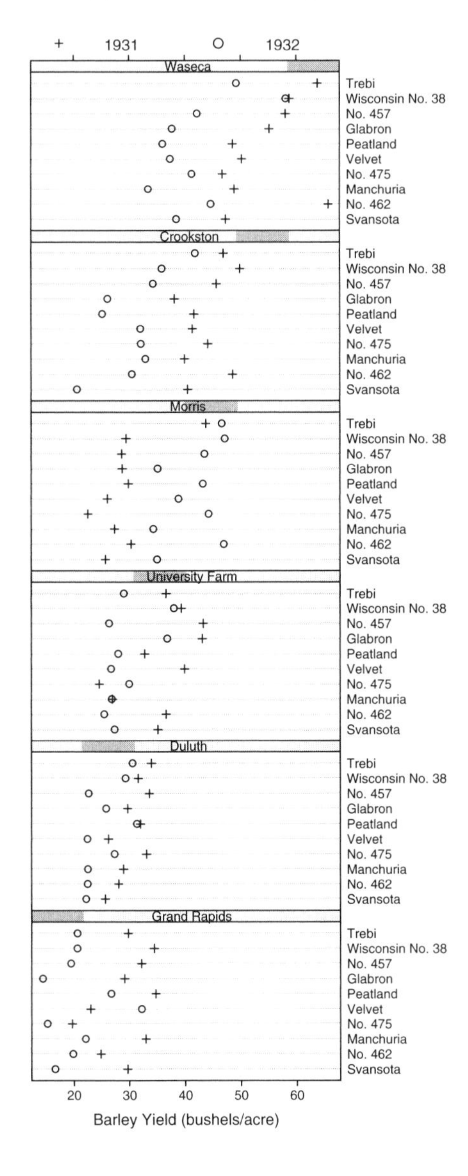

5.2 MORE THAN THREE VARIABLES

5.2.1 Superposed Data Sets

Many early statistical experiments were run on agricultural data. One of these studied the yield of barley. Ten varieties of barley were planted in six sites in Minnesota in 1931 and 1932. Yields were obtained for each combination of site, year, and variety. Therefore, there were 120 data points ($10 \times 6 \times 2$). Figure 5.4 shows these observations. Each panel contains a dot plot of the yield for each variety and each year for a specific site. Study this figure and comment on an interesting aspect of the data.

A study of Figure 5.4 shows that for five of the six sites, higher yields of barley were produced in 1931 than in 1932. However, in Morris, the reverse is true. Moreover, the amount by which 1932 exceeds 1931 in Morris is about the same as the amount by which 1931 exceeds 1932 in the other five sites. This anomaly was caused either by natural events such as unusual weather conditions, or by accidentally reversing the years. Interesting analyses to determine which actually happened are described by Cleveland (1993). The conclusion is that the data are in error; the years are reversed.

This data set appeared in a classic statistics book, *The Design of Experiments*, by R. A. Fisher (1971). Since publication made the data well known, a number of statisticians reanalyzed the data to demonstrate new methods. The error was discovered by applying trellis display after many other analyses over 60 years had failed to notice it. It shows the power of this visualization tool and the necessity of plotting data.

Fig. 5.5 Barley Data: Bar Chart

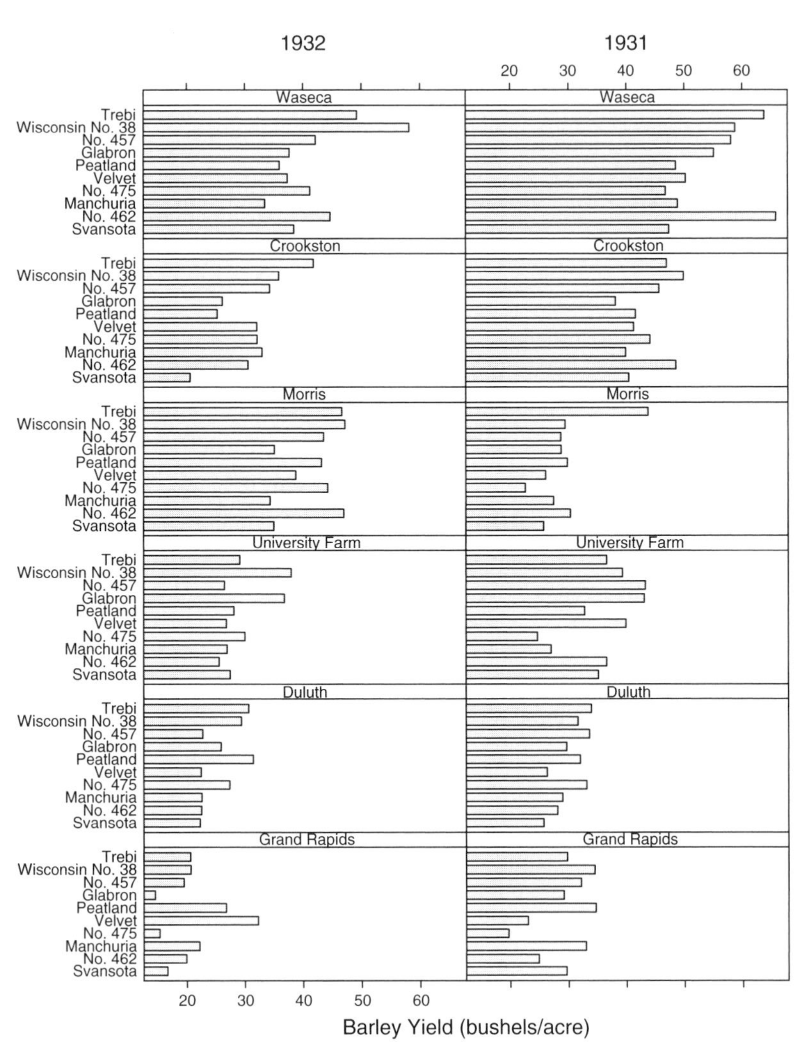

5.2.2 Trellis Multipanel Displays

We have used dot plots in a number of situations where bar charts are used customarily. In some cases we have also shown examples using charts available with commonly available programs. When we have only a few data points, it does not make much difference which chart we use. Figure 5.5 shows that bar charts become cluttered quickly and do not show the error at the Morris site as clearly. Figure 5.5 shows 120 values: six sites with 10 varieties of barley over two years. Many effective graphs show thousands of points. Using only common graphical packages is easy but limits what you are able to display.

Fig. 5.6 Energy Data: Scatterplot Matrix

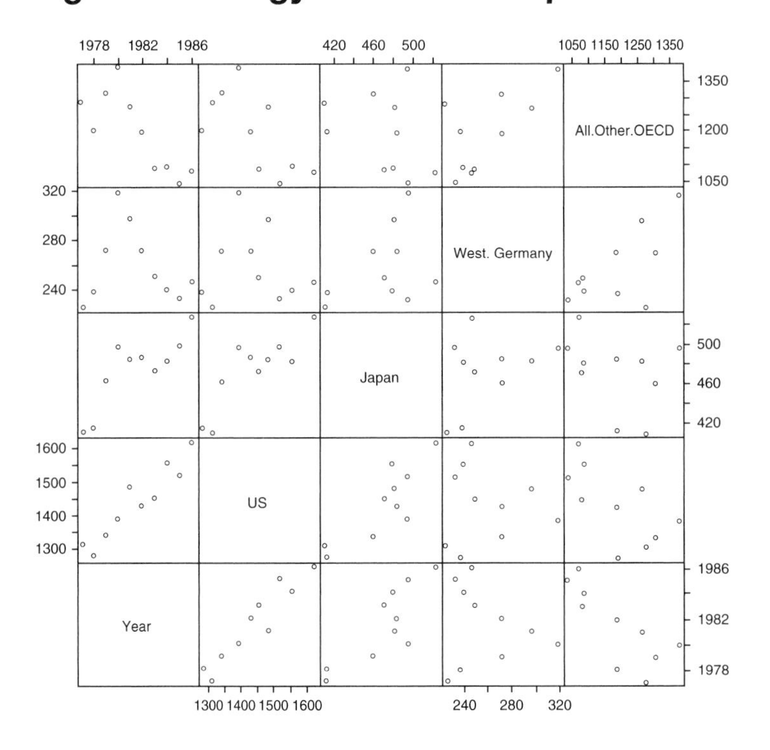

5.2.3 Scatterplot Matrices

A *scatterplot matrix* shows scatterplots for all pairs of more than two variables. For the energy data we have five variables: the year and the petroleum stocks for the United States, Japan, West Germany, and all other countries of the Organisation for Economic Co-operation and Development. All the plots in the bottom row of Figure 5.6 have years on the vertical axis, and the plots in the left column have years on the horizontal axis. The axis labels for all the plots in the first column appear on the top of the column; those for the second column appear at the bottom. They are used on alternate sides so that they don't run into one another. The plot in the top left corner shows the data in Figure 2.12. You may have wondered then how I knew that the petroleum stocks of all other OECD countries were decreasing with time. I scanned the scatterplot matrix and noticed that trend, which I couldn't see in the stacked bar chart.

Fig. 5.7 Reading a Scatterplot Matrix

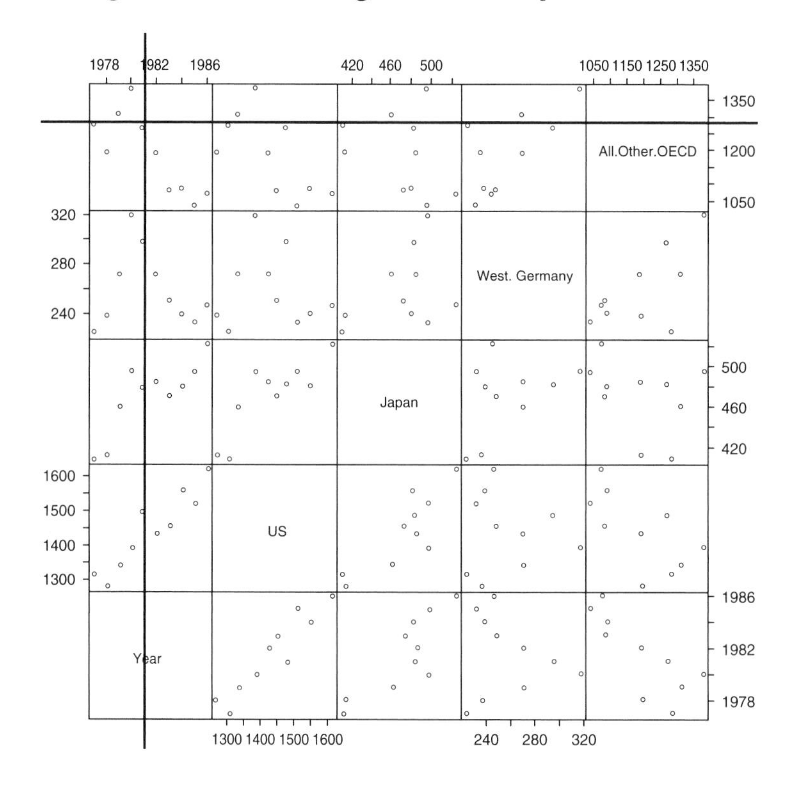

Scatterplot matrices are read much like the charts you see on road maps that give distances between cities. You pick out the row that has the first city you are interested in and then go across to the column with the second city. Their distance is at the intersection of the row and column. In our case we are interested in "all other OECD" and year in Figure 5.7. The horizontal line shows all plots with OECD on the vertical axis, and the vertical line shows plots with year on the horizontal axis. The intersection is the plot we are interested in. Notice that there is symmetry around the diagonal.[1] The bottom right shows year on the vertical axis with OECD on the horizontal, and the top left has OECD on the vertical axis with year on the horizontal.

[1] Heiberger and Holland (2004) explain why it is preferable to have the diagonal go from the lower left to the upper right in scatterplot matrices.

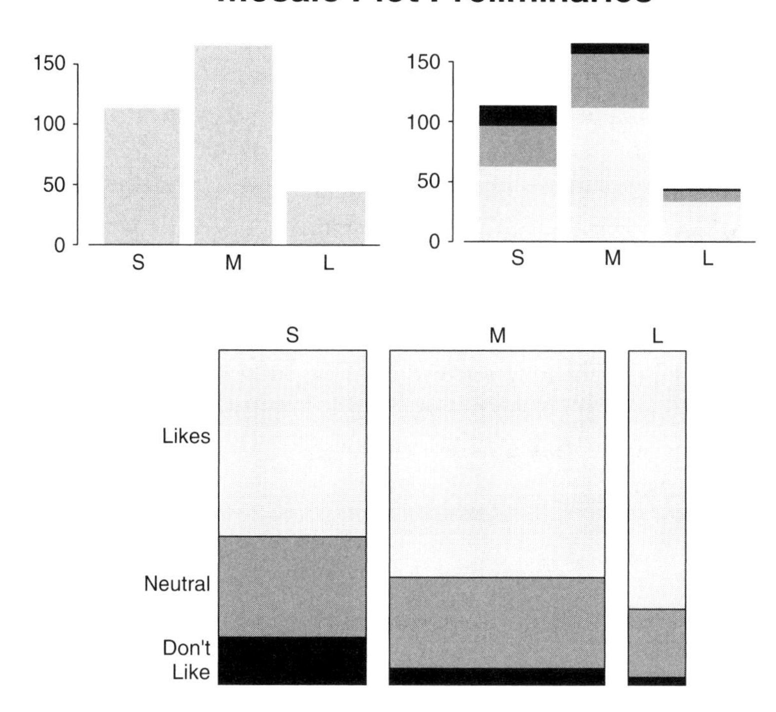
Fig. 5.8 Monterey Bay Aquarium Data: Mosaic Plot Preliminaries
150
100
50
0
S
M
L
150
100
50
0
S
M
L
S
M
L
Likes
Neutral
Don't
Like

5.2.4 Mosaic Plots

The Monterey Bay Aquarium (MBA) collected data on the length of time that visitors stayed at an exhibition as well as visitor ratings on a scale of 1 to 10 (with 10 being excellent), whether the visitor was in a group of adults or included persons under 18,[2] and whether this was a first or repeat visit to the aquarium (Monterey Bay Aquarium, 2001). The top left plot in Figure 5.8 shows a bar plot of the time visitors spent at this exhibition. The left column represents visitors who stayed a short time S (1 through 15 minutes), the middle column a medium stay M (16 through 30 minutes), and the right column a long stay L (31 minutes or longer). The top right plot adds the ratings, making this a stacked bar chart. The shading represents the ratings; the lightest shade of gray is 8 to 10 (likes), the middle shade is 6 to 7 (neutral), and the dark shade is 1 to 5 (don't like).

The bottom section of Figure 5.8 shows the same information in a *spine plot*, a chart where the width of the bars vary and all the heights are 100%. In a stacked bar chart the heights vary and the widths are the same. It is easier to compare the percentages of the ratings in a spline chart.

'divided bar'

[2] Groups with persons under 18 were labeled as families.

Fig. 5.9 MBA Data: Another Variable Added

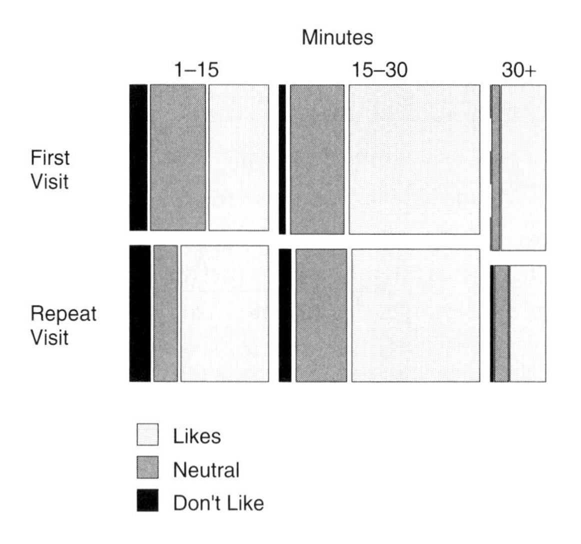

Figure 5.9 adds information as to whether the visitors are first-time or repeat visitors. Note that a higher percentage of those who stayed more than 30 minutes were first-time visitors than those who stayed shorter times.

Fig. 5.10 MBA Data: Mosaic Plot

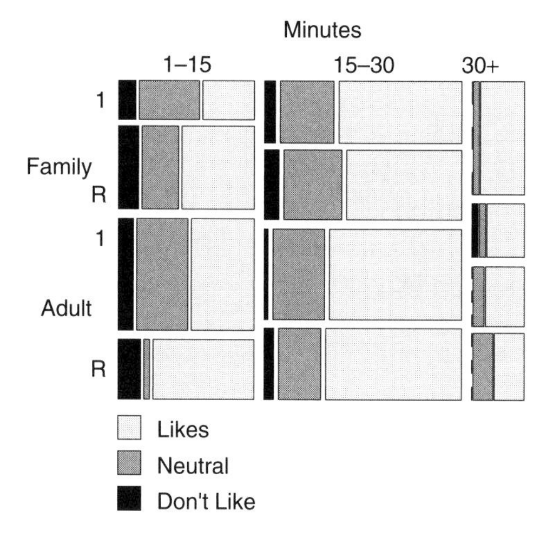

Figure 5.10 is a mosaic plot of the MBA data with all four variables. The numeral 1 stands for first-time visitors and R stands for repeat visitors. This figure includes the family or adult classification. A study of the plot shows that more visitors stay between 15 and 30 minutes than the other time groups, that ratings of 8 to 10 are the most common, and that only a small percentage of visitors give a rating of 1 to 5. The variables are not independent. A much smaller percentage of visitors who stayed a long time (over 30 minutes) gave ratings below 6. For visitors with short stays (1 to 15 minutes), repeat visitors gave more high ratings than did first-time visitors. This is not true for visitors who stayed over 15 minutes for this exhibition.

Fig. 5.11 Soybean Data: Linked Micromaps

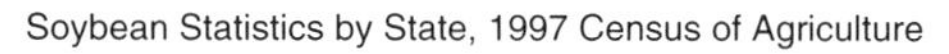

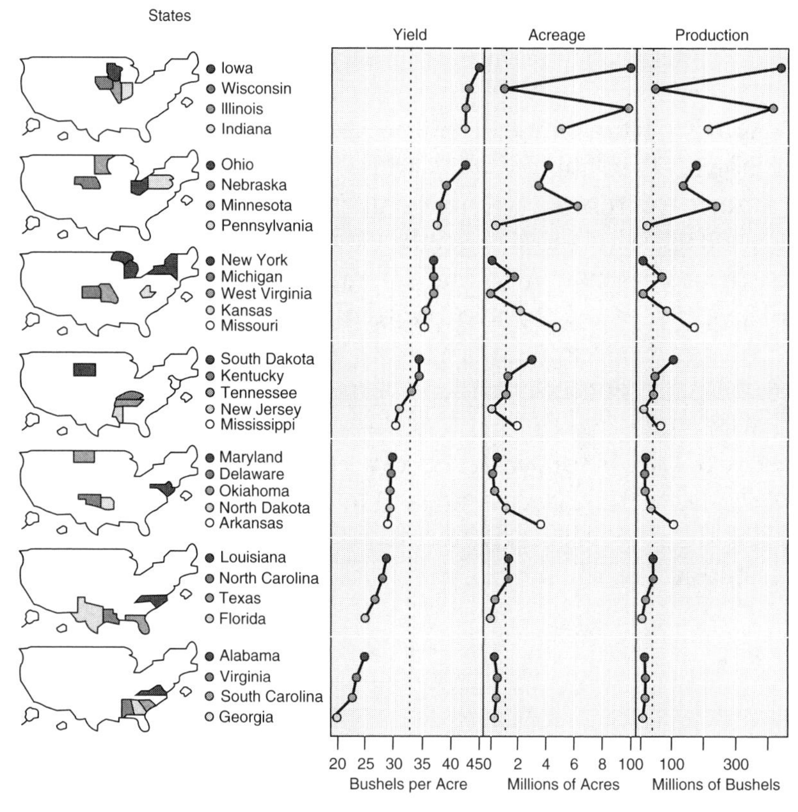

5.2.5 Linked Micromaps

Linked micromap (LM) *plots* are useful for displaying geographically referenced statistical data (i.e., data for areas such as states, counties, or economic regions) (Carr and Pierson, 1996). This linked micromap appears on the Web site[3] of the U.S. Department of Agriculture - National Agricultural Statistics Service. It shows the yield of soybean crops in bushels per acre as well as the acreage and production of these crops for the 31 states where soybeans were planted. The data are ordered by the yield. The vertical dotted lines represent the median of each variable.

[3] *http://www.nass.usda.gov/research/gmsoyyap.htm*

The data are broken up into groups of around five geographic areas. First, the four areas with the highest yield are plotted, with each area in the group a different shade. These dots are linked to a U.S. map that highlights the corresponding states in the same shades. Then this process is repeated with the group of states with the next highest yield.

The maps show the regions with high and low yields of soybeans, suggesting a trend of decreasing yields as we move from the midwest to the southeast. It is clear that the variables are correlated, but there are unusual cases. For example, Wisconsin produces a high yield with much less acreage than do the other states with high yields for the year shown. On the other hand, Minnesota, Missouri, and Arkansas use more acreage for their yields. Acreage and production are highly correlated.

LM plots replace *choropleth maps*, a common way of displaying statistical data for geographic regions. This example of a choropleth map is from Carr (1993):

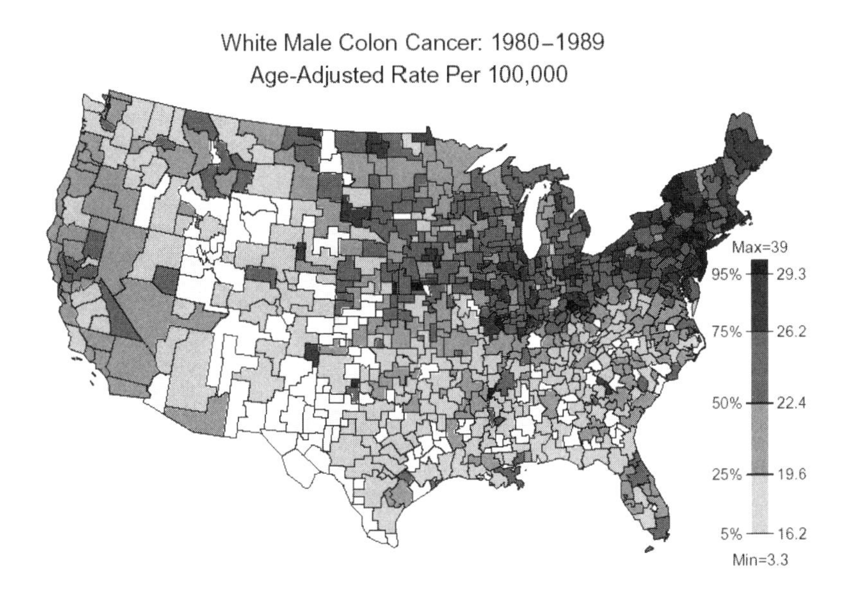

Choropleth maps have limitations (Symanzik and Carr, submitted, or Harris, 1996, p. 361) that LM plots correct.

Fig. 5.12 Iris Data: Parallel Coordinate Plots

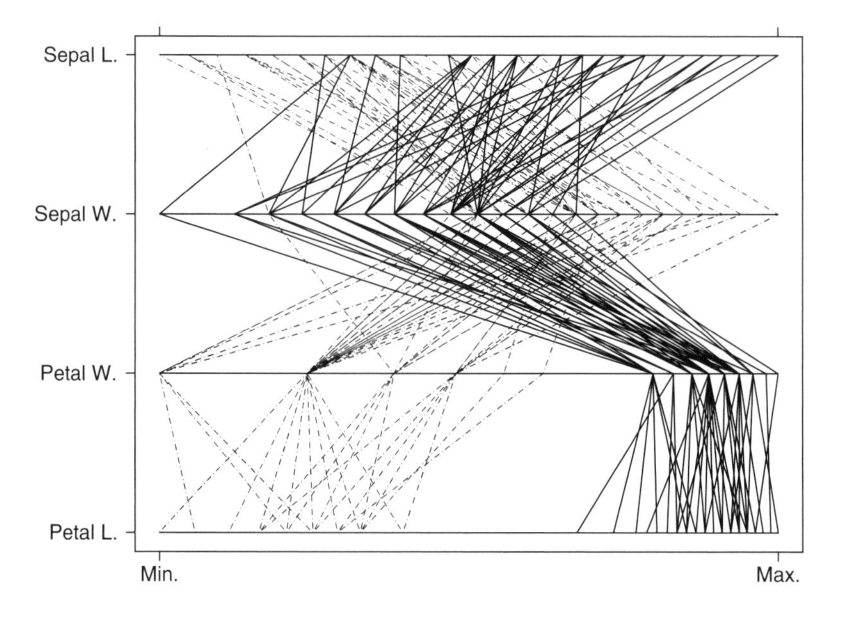

5.2.6 Parallel Coordinate Plots

Parallel coordinate plots are often used on large data sets to classify data into subgroups. Fifty irises from each of three varieties were sampled and measurements taken on their petal length and width and their sepal[4] length and width. This is another famous data set that was analyzed by R. A. Fisher in the 1930s (Fisher, 1936). Figure 5.12 shows two of these varieties: one shown by a solid line and the other by dashed lines. The variety shown with solid lines has a narrower sepal width and greater sepal length than those of the variety displayed with dotted lines. Also, the variety with the lower sepal width has larger petal lengths and widths.

[4] A sepal is one of the outer parts of a flower. It is usually green.

Fig. 5.13 Reading a Parallel Coordinate Plot

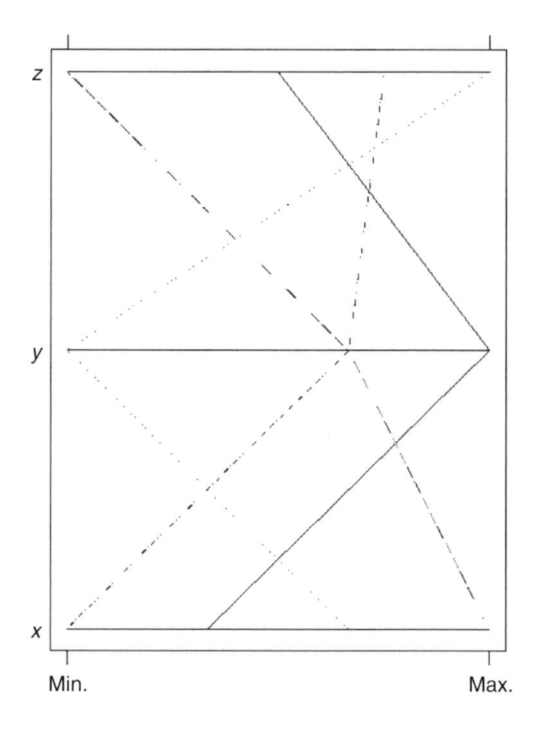

In a three-dimensional graph drawn with Cartesian coordinates, the x, y, and z axes are perpendicular to one another. To plot $x = 2$, $y = 4$, and $z = 3$, we move 2 units in the x direction, then 4 units in the y direction, and then 3 units in the z direction (usually, up). On the other hand, with parallel coordinates the axes are parallel to one another. The variables may have different units, so the scale for each variable ranges from its minimum value to its maximum value. Figure 5.13 plots the four observations (2,4,3), (3,1,4), (1,3,3.5), and (4,3,2). The range of x is from 1 to 4, y is from 1 to 4, and z is from 2 to 4. The solid black line represents the first observation, with $x = 2$, $y = 4$, and $z = 3$. Try to associate the three types of dotted lines with the second, third, and fourth observations.

Fig. 5.14 Nightingale Data

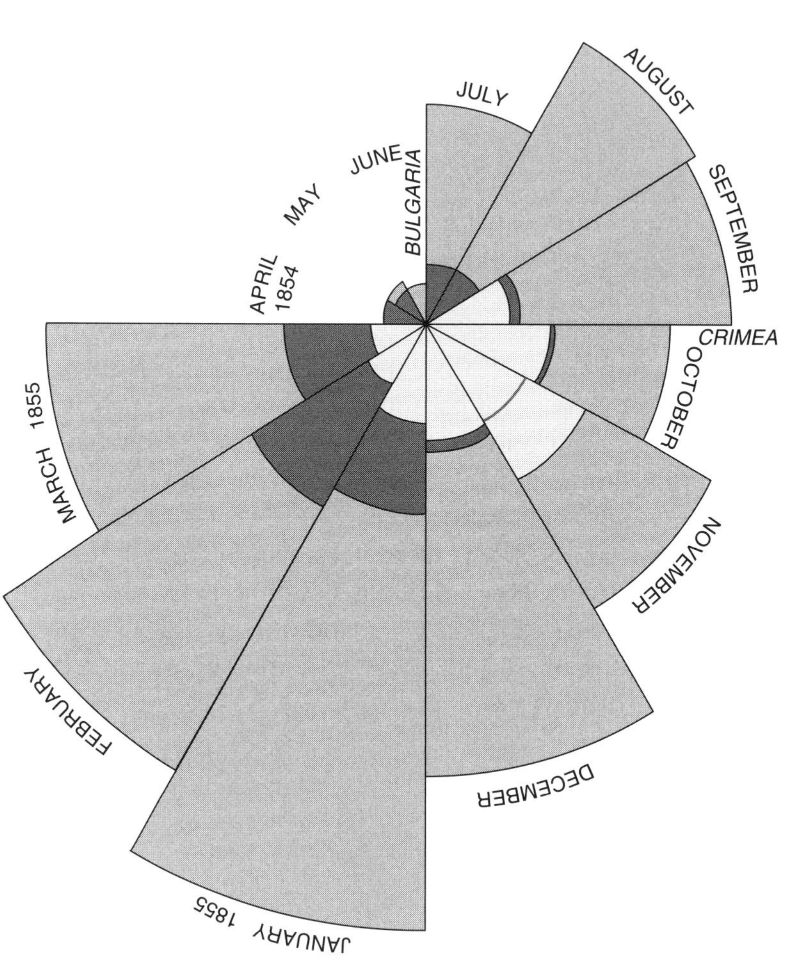

5.2.7 Nightingale Rose

Florence Nightingale introduced plots such as that of Figure 5.14 in 1858. I've seen them called *Nightingale rose plots, Nightingale coxcomb plots,*[5] *wedges graphs, radial area plots*, and *Nightingale's pie charts*. A pie chart has a fixed radius and varies the angle. But we learned that angles are hard to judge. These plots have fixed angles and vary the radius: The radius varies with the square root of the data so that we perceive areas accurately. In Figure 5.14 the wedges represent months. The lightest-colored layer closest to the center in most wedges represents deaths by wounds (red in the original), the middle shade represents preventable disease (blue in the original), and the dark shade represents other causes of death (black in the original). Each wedge is measured from the center of the circle; they are not stacked.

Crash – fatality: pdo, injury by month.

[5] See *http://www.florence-nightingale.co.uk/small.htm* for the reason that the term *coxcomb plot* is incorrect here.

For most months the mortality rate for all other causes exceeded that for wounds, but in November 1854 the rate for wounds was greater than for all other causes. Therefore, the November value for all other causes would be hidden without the line showing its value. The caption includes, the words: "The black line across the red triangle in Nov. 1854 marks the boundary of the deaths from all other causes during the month." These plots clearly show that many more of the combatants died from disease than from wounds and played a large role in Nightingale's arguments to improve sanitary conditions for the British forces during the Crimean war.

I've seen comments on newsgroups asking why more books don't describe these plots. Part of the problem may be that not many software packages include them. Versions of this figure appear on the Web that were created with SAS and CorelDraw. Code is available to use with other software. I found code for S-Plus by searching on *circular statistics*.

We decode the information in Figure 5.14 by visualizing the areas of the wedges. Suppose that we have two numbers, one double the size of the other, and we display them using areas of squares or circles. Some designers make the side of a square or radius of a circle proportional to the values, whereas other designers make the square root of the side or radius proportional to the values so that we perceive the area correctly. Florence Nightingale drew this figure using square roots so that we perceive the areas accurately. However, it is so common to use the side of the square or radius of the circle that the reader does not know whether or not a figure showing areas was drawn correctly. Chapter 6 contains several examples of graphs with this problem. Therefore, it is best to avoid using areas for changes in one dimension.

Fig. 5.15 Nightingale Data: Trellis Plot

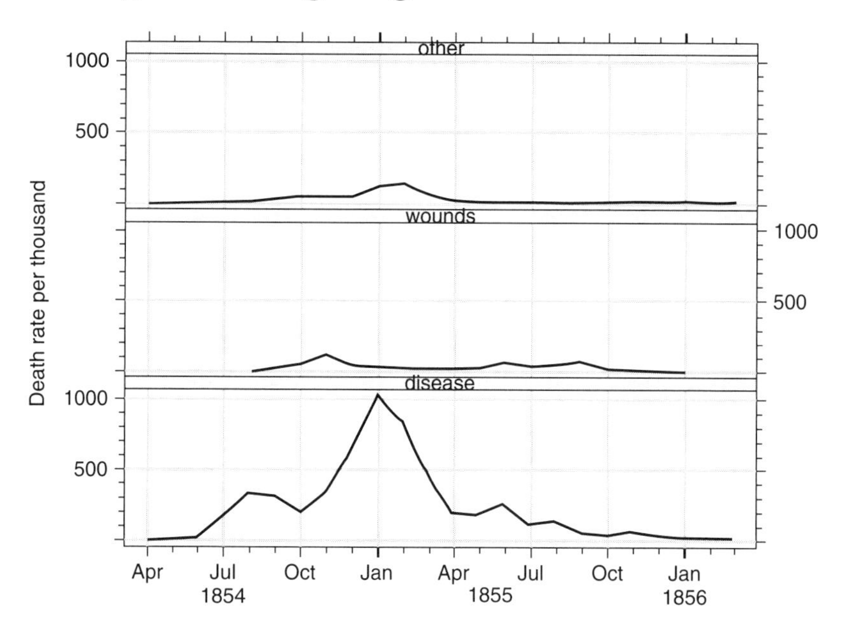

The data for April 1854 through March 1855 in Figure 5.15 are the same as those that appear in the Nightingale rose plot in Figure 5.14 The right side of Figure 5.15 shows the death rates for the following year, showing the dramatic reduction after sanitary measures were implemented. Both the Nightingale plot and this plot show clearly that the death rate from disease far exceeded that from wounds and other causes. Whereas Florence Nightingale's graph gives an impression of the relative sizes of the wedges, this trellis plot allows us to determine the actual death rates.

Fig. 5.16 DJIA: High–Low–Open–Close Plots

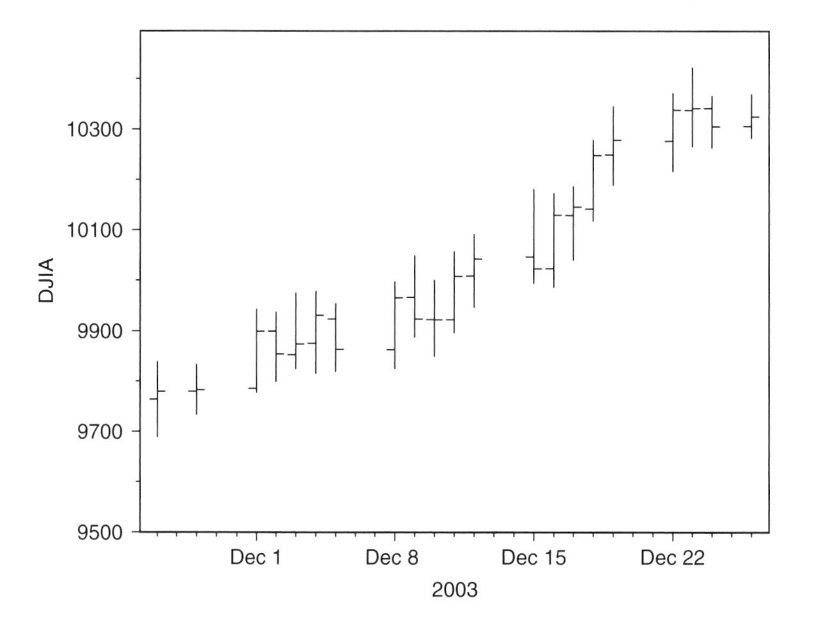

5.2.8 Financial Plot

In addition to the general-purpose graphs discussed, there are many specialized graphs with application to specific fields. Although I do not discuss most of these, the principles in the next two chapters should guide your creating effective graphs in any field. Figure 5.16, useful in finance, shows the highs, lows, and opening and closing values of the Dow Jones Industrial Average (DJIA) from Thanksgiving through December 26, 2003. The vertical lines go from low to high, with the left horizontal lines representing the opening values and the right horizontal lines representing the closing values.

5.3 COMMENTS

In this chapter some methods have been presented for displaying graphs with more than two variables that can be created using commercial or open-source statistical software. The problem with using many software packages is that they limit your thinking on how to present your particular data set. They contain a list of graph types from which to choose. Better ways to present your data may not be included as types on their list. Some software lets you modify its code, giving you more control. Figure 8.9 provides an example of a customized plot.

There are examples of effective multivariate plots all around you. Look at the weather chart in any copy of the *New York Times*. It shows the forecast high and the forecast low for today and the next four days, the actual high and the actual low for the preceding five days, the normal highs and lows for all 10 days, and the record highs and lows for all 10 days in a simple, uncluttered chart.

A graph that many consider to be the best chart ever drawn is *Napoleon's March to Moscow* by Minard, dated 1869 (Tufte, 2001, pp. 40–41). This graphic moves over space and time as it shows the fate of Napoleon's army in Russia while plotting six variables. You can see it at

http://www.math.yorku.ca/SCS/Gallery/minard/orig.gif

SUMMARY

There are a number of techniques for displaying more than two variables on two-dimensional paper or screens. The multipanel displays of trellis graphics are particularly useful and can be used to replace stacked or grouped bar charts. Charts for special purposes include mosaic plots for categorical data; high-low-open-close for financial data; and linked micromaps for geographically referenced data.

6 General Principles for Creating Effective Graphs

Cleveland (1994) presents principles for creating effective graphs in science and technology. In this chapter and Chapter 7, Cleveland's principles are adapted to nonscientific examples. In addition, I have added principles that are needed for business and other nonscientific graphs, as these graphs have different problems from those of scientific graphs.

The first set of principles ensures that the reader clearly sees what is graphed. This won't happen if extraneous information masks the data or data points overlap. Visual clarity is important not only for the data but for other graphical elements, such as tick marks, labels, and axes, as well. The

Creating More Effective Graphs, by Naomi B. Robbins
ISBN 0-471-27402-X Copyright © 2005 John Wiley & Sons, Inc.

next set ensures that the reader clearly understands what is graphed. This requires that the data be drawn to scale and that there be clear labels and explanations. Some principles on general strategy follow. Scales have a major influence on our interpretation of graphs. Principles related to scales are given in Chapter 7.

The examples of graphs violating these principles are real examples of graphs I have seen wherever I have obtained permission to use these examples. However, when my requests for permission went unanswered, I have shown the problems with simulated examples.

Fig. 6.1 Terminology

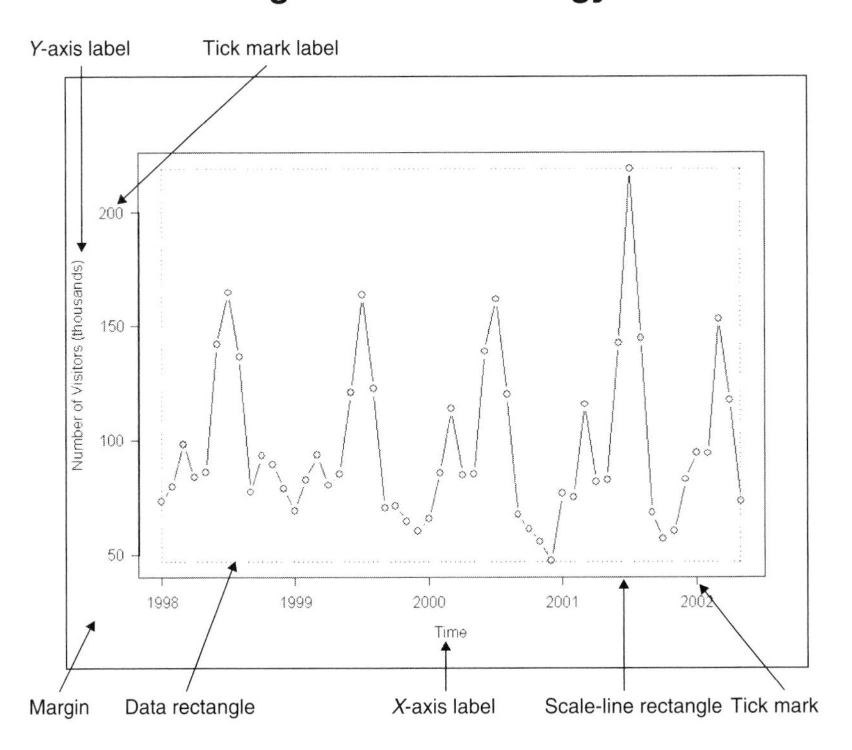

6.1 TERMINOLOGY

Figure 6.1 defines the areas of a graph that I refer to in this chapter. Some are imaginary rectangles, which define the space that we have to work with in a graph. The dotted rectangle, called the *data rectangle*, just encloses the data points. The scale lines are the rulers along which we measure the data; the scale-line rectangle is formed by the horizontal or x axis and the vertical or y axis. Finally, the outermost rectangle includes the margin, which contains the information we need to understand the graph: title, axis or scale labels, tick marks, tick mark labels, and usually, a caption. A legend or key may be included when there is more than one set of data.

Fig. 6.2 Association of Research Libraries Data

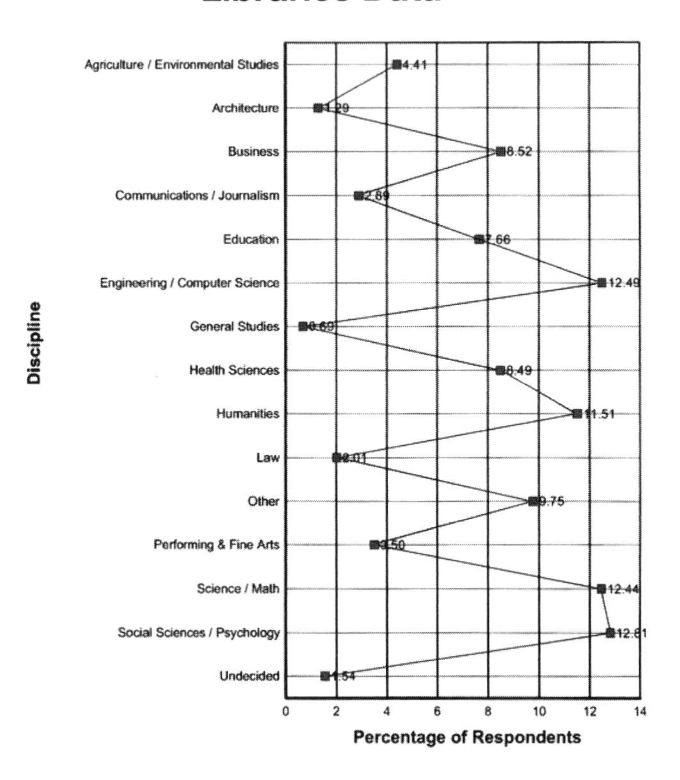

Source: ARL Notebook 2002 LibQUAL + (tm) Results (Washington, DC: Association of Research Libraries, 2002). 4.1.2: Respondents by Discipline for 4-year Institution—All User Groups (Includes Library Staff)[table], p. 26. Retrieved January 20, 2003 from

<http://www.libqual.org/documents/admin/ARLNotebook111.pdf>
For more information on the ARL Statistics and Measurement Program, see *<http://www.arl.org/stats/>*

6.2 VISUAL CLARITY

6.2.1 Clarity of Data

Make the data stand out. Avoid superfluity[1]**.**

If I were asked to summarize the message of this book in five words, I would say "Make the data stand out." If limited to three, my words would be "Emphasize the data." This book presents graph forms, advice, and commentary on how to implement these dictums. When you draw a graph or look at a graph drawn by others, ask yourself "What is the first thing I notice?" The answer should be "the data."

What do you see when you look at Figure 6.2?

[1] Any principle that is printed in ***bold italic*** type is a direct quote from Cleveland (1994); those not italicized are ones I've added.

Fig. 6.3 ARL Data: Data Stand Out Better

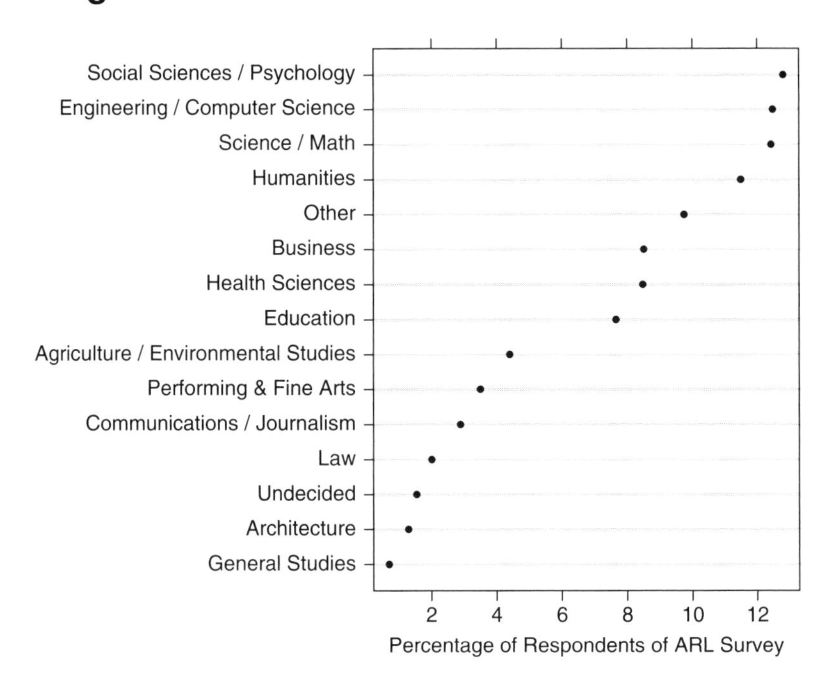

I see lines. I see grid lines and I see lines connecting the data points. Note that if we translated the alphabetically ordered labels into another language, the visual impression of the figure would be completely different, since the disciplines would be reordered and the lines connecting the points would have a different shape.

Far too often when I ask myself what I see, my answer is that I see prominent grid lines or I see a string of zeros in the labels. The same data stand out better in Figure 6.3. In addition to eliminating some lines and deemphasizing others, the disciplines are reordered by the percentage of respondents. This helps the reader to understand the data.

Fig. 6.4 Data That Are Difficult to See

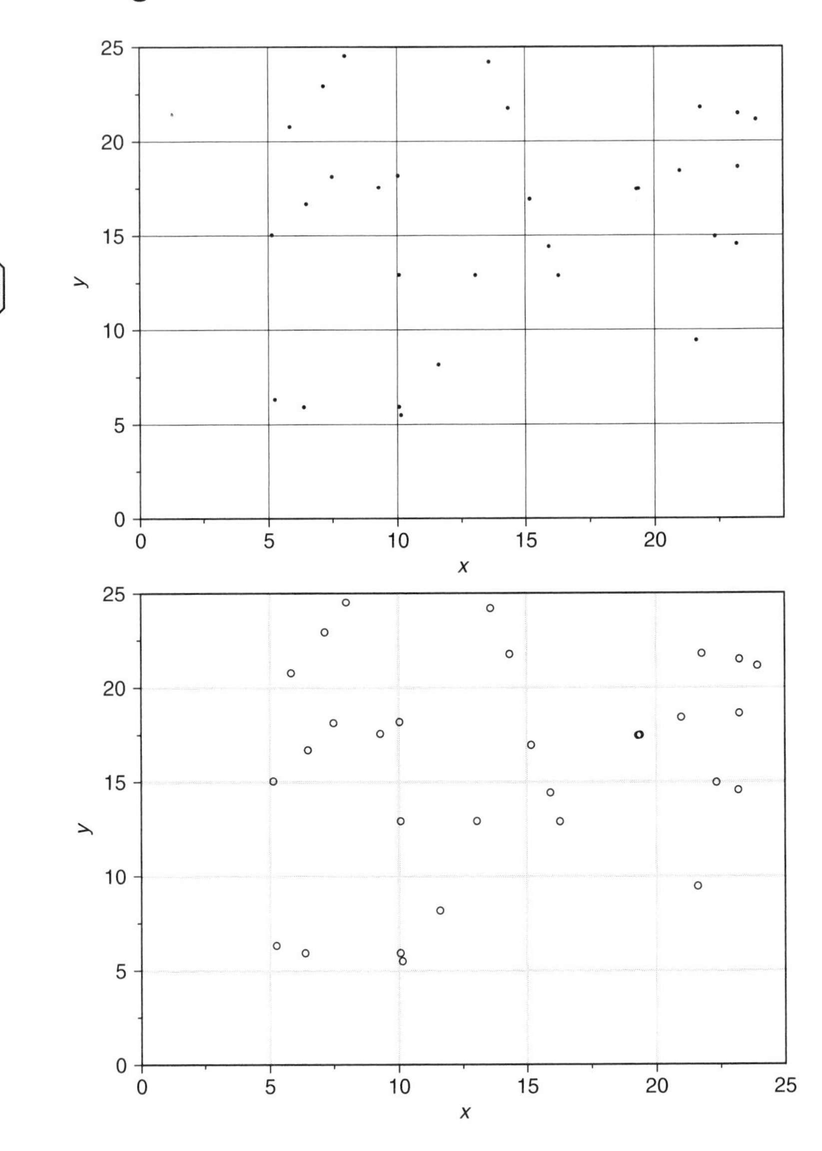

Use visually prominent graphical elements to show the data.

At the top of Figure 6.4, the data symbols are too small and difficult to see; the grid lines are more prominent than the data. A number of points are especially hard to see since they fall on the grid lines. In contrast, in the bottom figure the data are more prominent and the grid lines are deemphasized.

Fig. 6.5 Museum Exhibitions: Overlapping Data

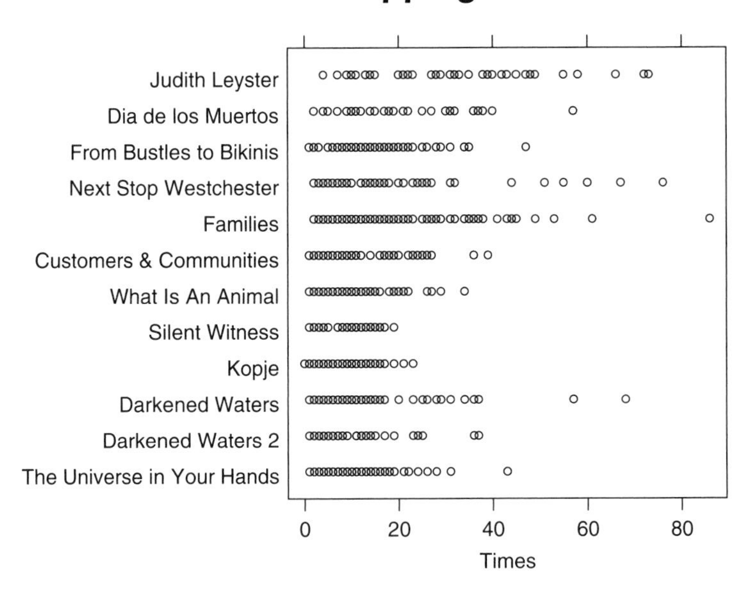

Overlapping plotting symbols must be visually distinguishable.

Figure 6.5 has many overlapping plotting symbols, due to duplicate values that occurred because the times were measured only to the nearest minute. As discussed in Chapter 4, jittering provides one solution to this problem. Another solution that works with a small number of overlapping plotting symbols is to alter the locations of points slightly to show all points. The caption should mention when this is done. A third possibility is to take logarithms to eliminate the overlap when the points don't have exactly the same values. Finally, it is easier to detect partial overlap when the plotting symbols are open circles than when they are filled circles.

Fig. 6.6 Superposed Data Sets

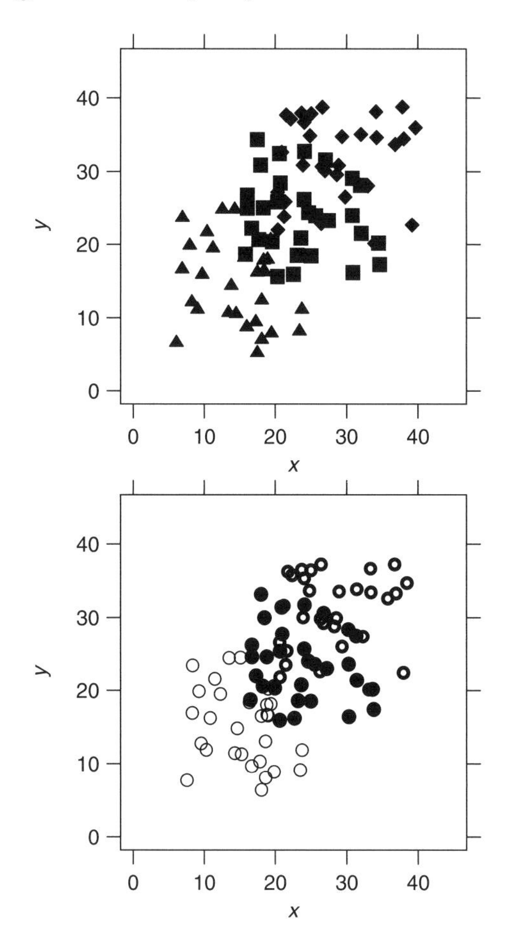

Superposed data sets must be readily visually assembled.

Three groups of data are plotted in Figure 6.6. Of course, a legend would be included with real data. The plotting symbols in the top plot are similar to one another, making it difficult to assemble the three sets of data. This job is much easier with the symbols used in the bottom plot. However, there is considerable overlap in these data sets which does not show up well with the solid data markers. These symbols work well when there is no overlap in the data. Color, if available, allows the best discrimination of data sets.

Fig. 6.7 Superposed Data Sets 2

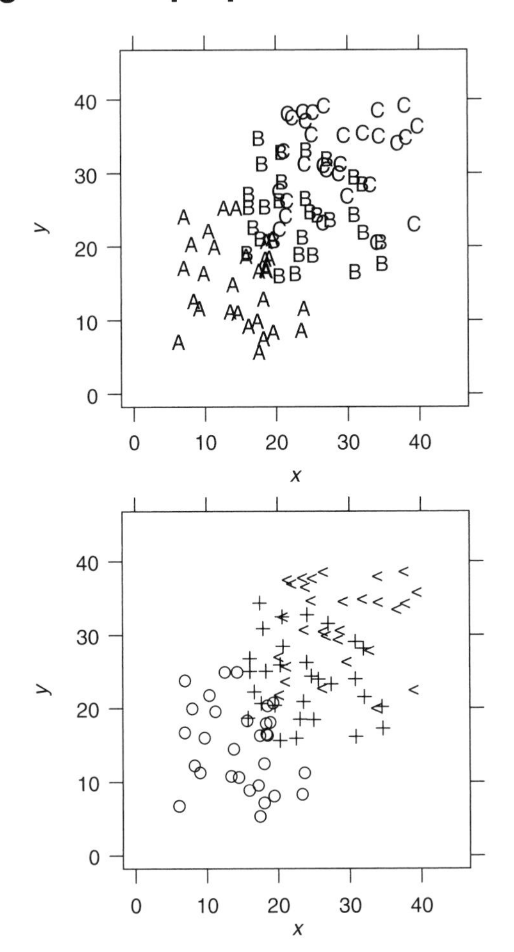

Using letters as symbols as in the top plot in Figure 6.7 eliminates the need for a key or legend and avoids the need for going back and forth to identify the data sets. However, it is more difficult to assemble the three data sets effectively than with some of the other combinations of data symbols.

The symbols in the bottom plot seem to work best for data sets with significant overlapping when color is not available. This choice of symbols is based on experimentation with texture perception (Cleveland, 1994, p. 238).

Fig. 6.8 Avoiding Superposed Data

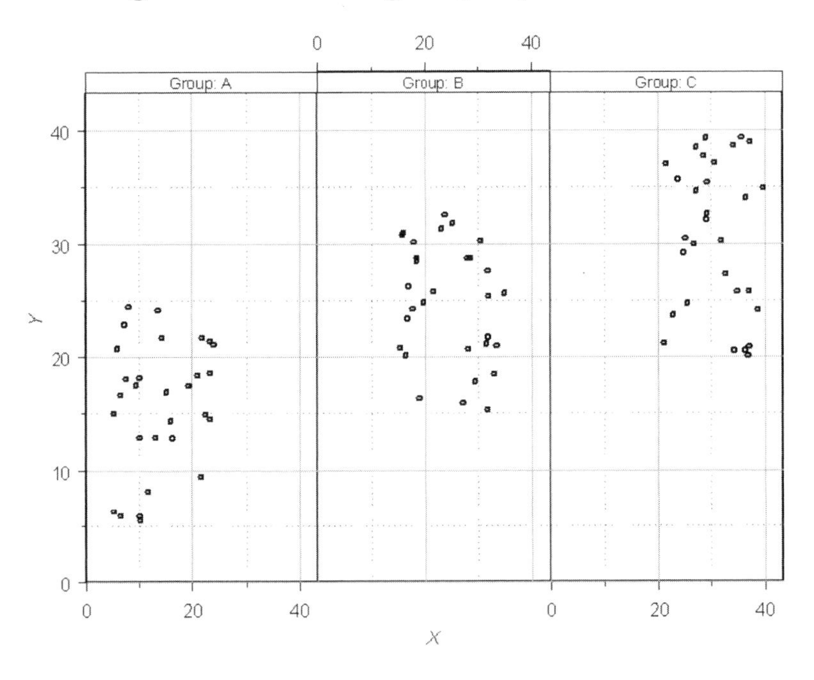

This trellis display of the data avoids the need to choose data symbols that can be distinguished easily. In addition to creating three panels, the data have been jittered in the y direction. Jittering both variables eliminated some overlapping symbols but caused other symbols to overlap. This presentation of one row with three columns facilitates comparisons of the y variable; three rows with one column would facilitate comparisons of the x variable. Although this presentation of the data makes separating the data sets more accurate, it makes comparisons in the horizontal direction less so. The grid lines help to make these comparisons.

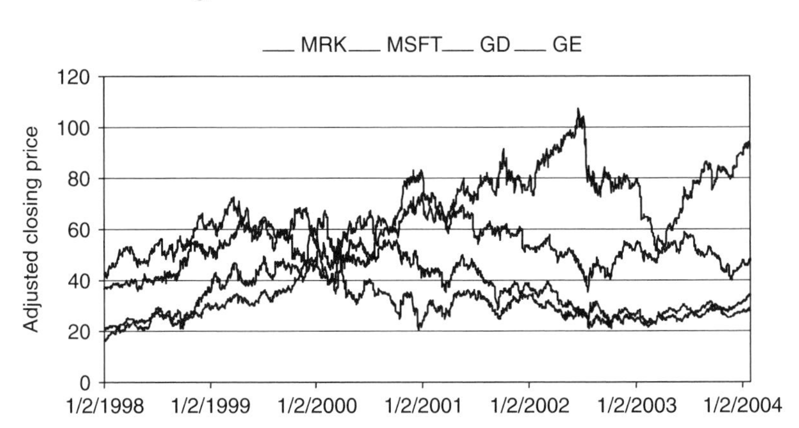

Lines must be distinguishable as well. How often have you seen charts that look like the mess shown in Figure 6.9, which shows the adjusted closing prices of Merck, Microsoft, General Dynamics, and General Electric from January 1, 1988 to February 1, 2004? It is impossible to follow any of the lines since they cross one another. Using different colors or line types (dotted, dashed, etc.) helps. Another way to aid readability is to have each line in a separate panel of a trellis display.

Fig. 6.10 Stock Market Data: Clarified

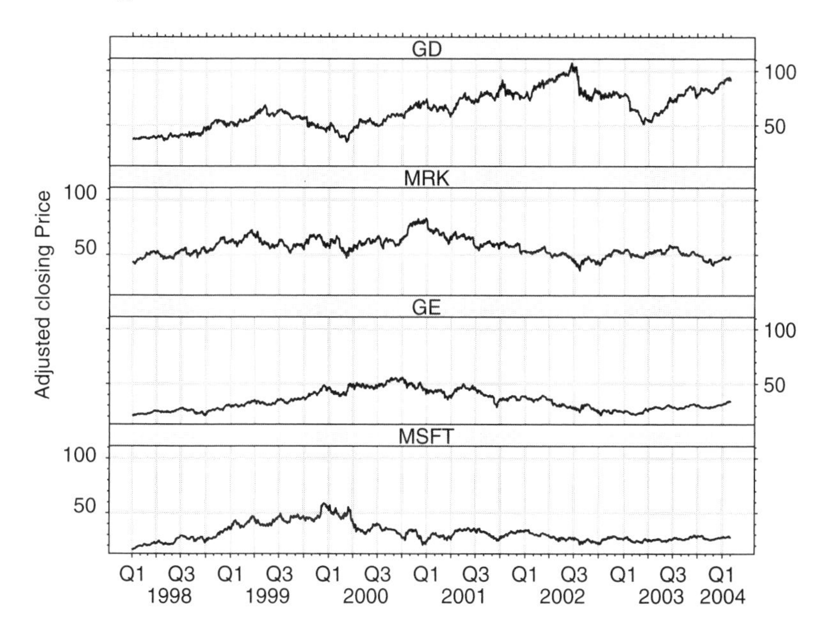

The stock market data are redrawn in Figure 6.10 with each stock in a separate panel.

Fig. 6.11 Marketing Data

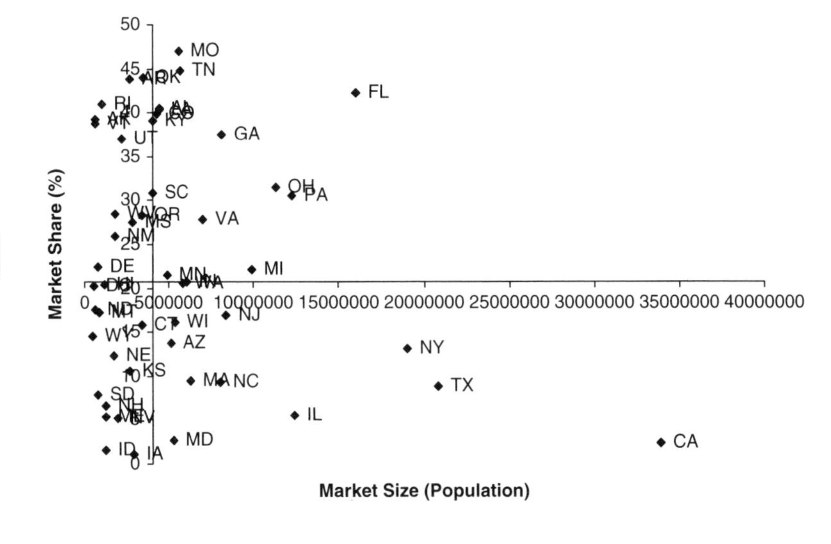

Do not clutter the interior of the scale-line rectangle.

Sometimes it is very difficult to design an uncluttered graph. For example, a marketing group had very specific requirements for a graph of market share versus market size (Figure 6.11). The axes had to cross at the medians of market size and market share so that the reader could tell at a glance if a given market was in the top or lower half of the markets. Also, all points had to be labeled.[2] The data and the tick mark labels obscure one another, and the figure is difficult to read. Including axis labels would cause even more clutter, as would using the full state names.

---

[2] Inserting data labels in Excel is described on page 339.

Fig. 6.12 Marketing Data: Less Cluttered Version

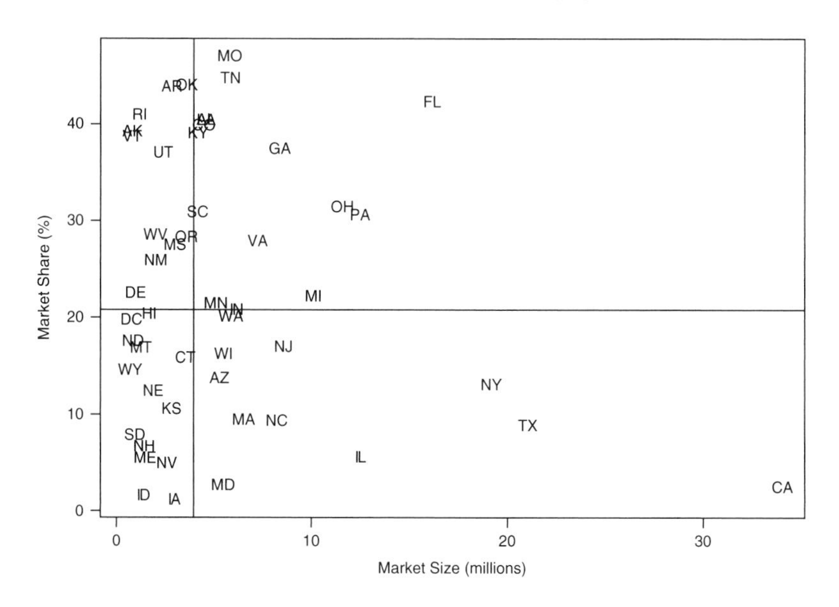

Figure 6.12 shows the same data with an attempt to reduce the clutter. The scale lines with their tick marks and labels were moved to the sides of the scale-line rectangle with reference lines at the medians to meet the requirement of being able to see whether a given market is in the top or bottom half. Rather than label the data points, state abbreviations were used as the plotting symbols. This eliminates the need for both points and labels. The population is given in millions to reduce the number of zeros for the *x*-axis tick mark labels.

Fig. 6.13 Inflation Data

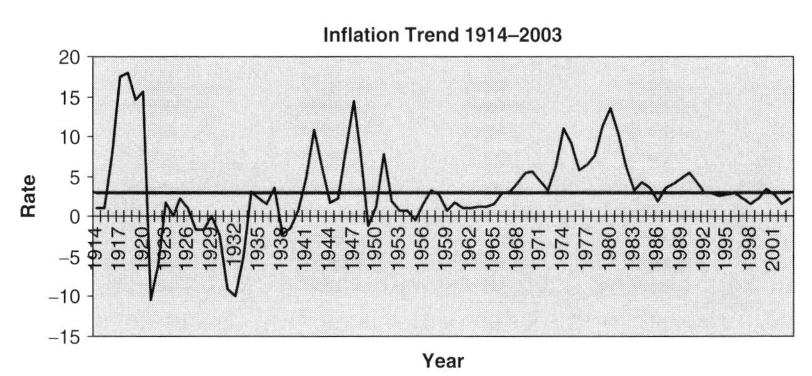

The horizontal line represents an inflation rate of three.

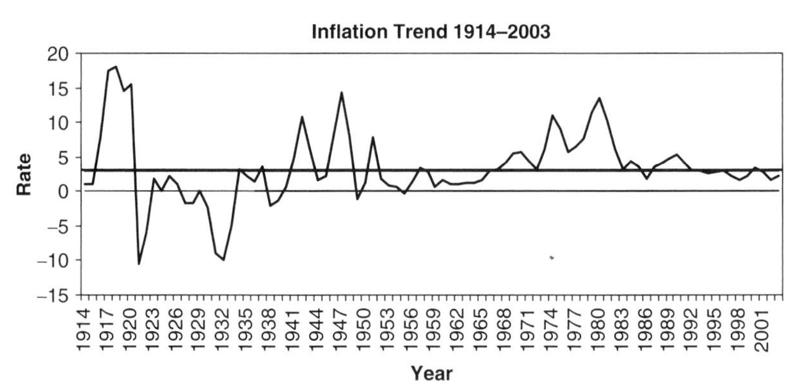

The horizontal line represents an inflation rate of three.

6.2.2 Clarity of Other Elements

Use a pair of scale lines for each variable. Make the data rectangle slightly smaller than the scale-line rectangle. Tick marks should point outward.

In the top plot in Figure 6.13, the horizontal scale line falls in the middle of the data. As a result, the tick marks and tick mark labels interfere with the data. This problem occurs often in charts drawn with Excel, since the scale positions shown here are the default in Excel. The solution is to move the horizontal scale away from the data as shown in the bottom plot. A similar example is included in Chapter 9 with instructions for performing these steps with Excel.

Fig. 6.14 Data That Hide Labels

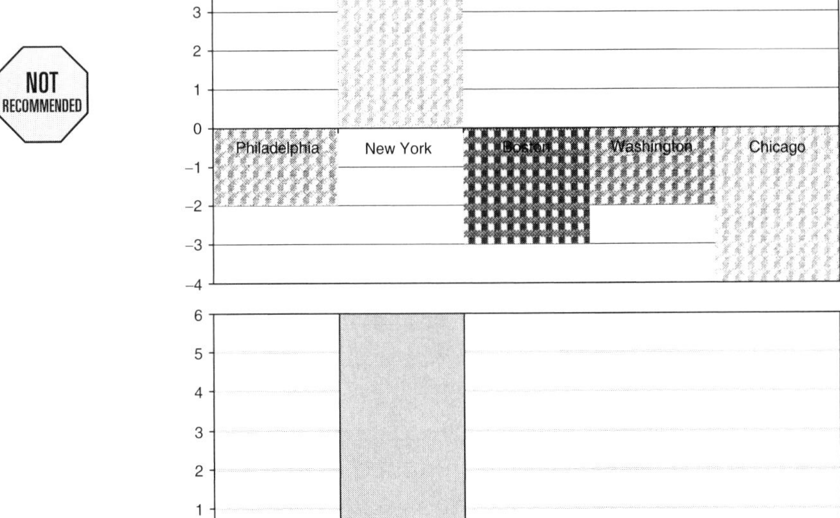

Figure 6.14 is another example of violating the principle that the data rectangle should be slightly smaller than the scale-line rectangle. Again, the scale line falls in the middle of the data. Although the data do not stand out in the top plot of Figure 6.13, the labels are still legible. That is not the case here. It is difficult to read some of the group labels since the patterns of the data mask the labels. The labels interfere with the data as well.

The problem is corrected in the bottom plot. I've also replaced the plaid patterns, as these only distract from the data.

Fig. 6.15 Tick Marks and Tick Mark Labels

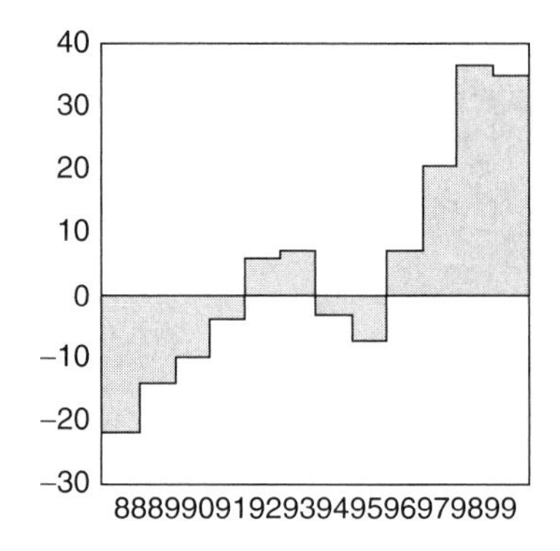

Do not overdo the number of tick marks.

Do not overdo the number of tick mark labels.

This is a very common error. Many people who prepare graphs label every tick mark or, as in this case, every step or bar. The result may look like the label on the horizontal axis of Figure 6.15 (Hager and Scheiber, 1997). The label looks like it is one long number instead of the years 1988–1999 written in yy format.

The solution is simple: Use only as many tick marks as needed to understand the data, and don't necessarily label all of them. The graph would be much easier to read if only the even years were labeled. Readers would easily grasp the missing years.

Another problem with this graph is the lack of tick marks corresponding to the *y*-axis labels, which makes estimating values less accurate.

Fig. 6.16 Exhibit Label Data

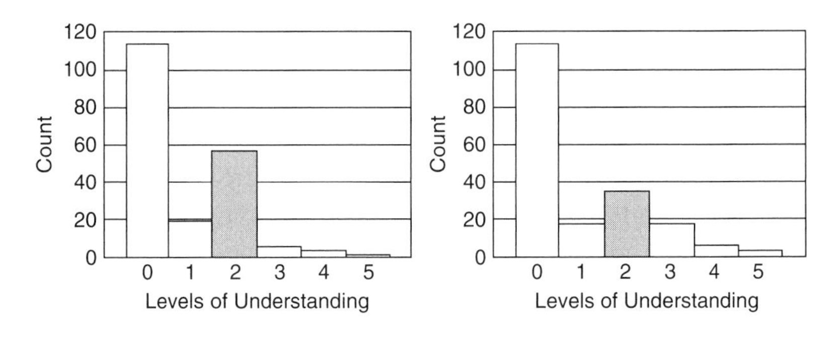

Deemphasize grid lines and distinguish grid lines from data.

The pair of charts in Figure 6.16 shows the distribution of the levels of understanding in a museum exhibition before (left figure) and after (right figure) revisions to the exhibit labels. We see that there were more scores of 3, 4, and 5 after the revisions. Look at the level of understanding of 1 in the left chart. Is the value 20 or is it zero? It is difficult to determine since there is a grid line at 20. This problem would be avoided if the grid lines were dotted or colored a light shade of gray.

Fig. 6.17 Stock Market Data: Visual Reference Grids

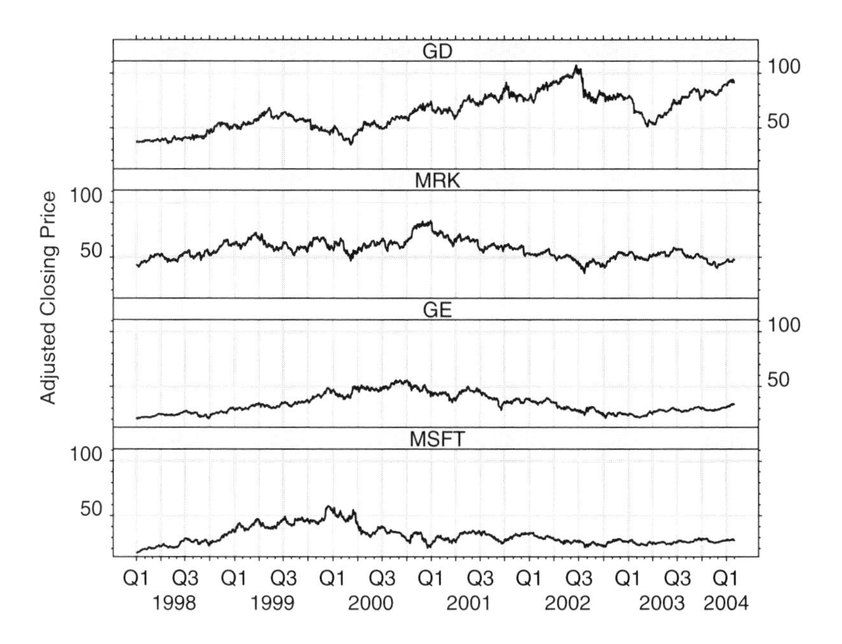

We have seen that many grid lines are so prominent that they distract from the data and also that grid lines can mask the data. For these reasons, some authors (e.g., Bigwood and Spore, 2003) recommend eliminating grid lines altogether. They claim that the tick marks can be used to read the data. However, in some cases I find unobtrusive grid lines helpful for estimating data values. A more important use for grid lines is to serve as visual reference grids when comparing data on different panels. Look at Figure 6.17 and notice how the reference grids help to compare the relative locations on the various panels. Judgment is needed on a case-to-case basis to determine if grid lines are helpful or clutter the graph.

Fig. 6.18 Zoo Exhibit Data

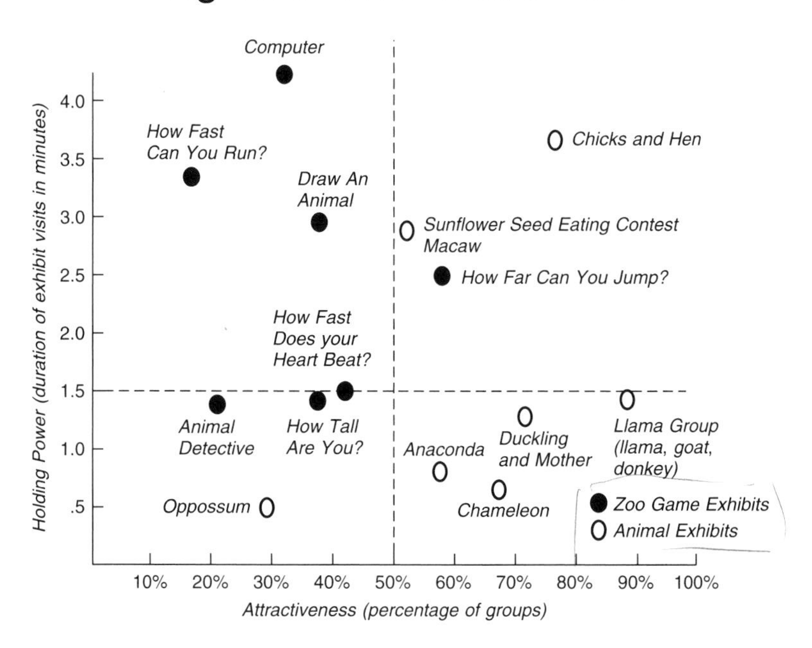

Avoid putting notes and keys inside the scale-line rectangle. Put a key outside and put notes in the caption or in the text.

A natural reaction when seeing Figure 6.18 (Rosenfeld, 1982) is to ask: What do the open and closed symbols represent? There is a key in the bottom right portion of the graph showing that the closed circles represent zoo games and the open circles represent animal exhibits, but the key is difficult to distinguish from the data. This would be avoided if the key were outside the scale-line rectangle. If space absolutely does not allow the key to be outside and there is enough white space within the scale-line rectangle for the key, distinguish it from the data by inserting it in a box.

Fig. 6.19 Barley Data

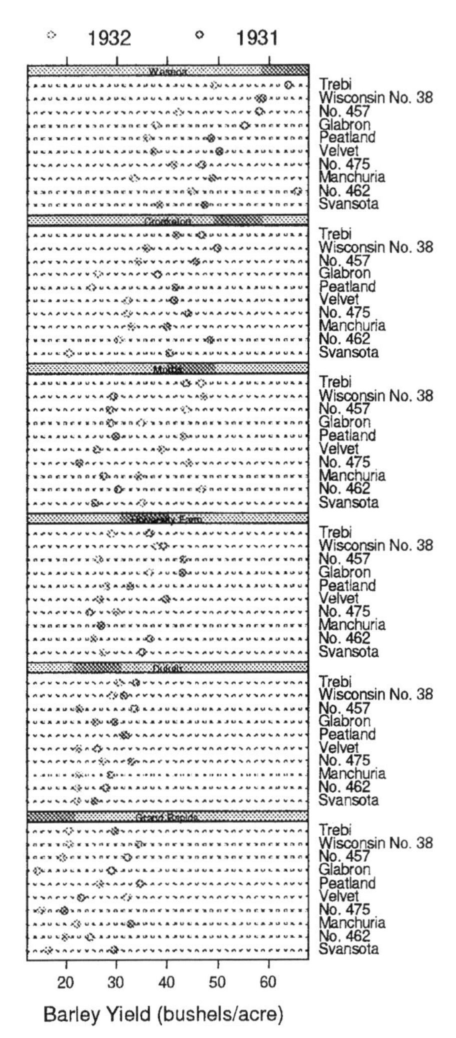

Visual clarity must be preserved under reduction and reproduction.

Many of the most carefully drawn graphs become useless after reproduction or reduction. Shades of gray that are beautiful if printed individually sometimes become indistinguishable when copied. Details are often lost when faxed. I didn't have to go far to find an example of this principle. One time I gave a paper at a professional meeting. The instructions said that authors could submit two versions of their papers for the *Proceedings:* one in black and white for the paper copy and the other in color for the CD version. I carefully prepared two versions, labeled them properly, and submitted them. The figure was the trellis display shown in Figure 6.19. (Robbins, 1999). The black-and-white version used different symbols for the years;

the color version used two colors of circles for the two years. Imagine my dismay when I saw that they had reproduced the colored version in black and white to produce the paper version. It was impossible to distinguish the two years. My presentation had emphasized the importance of distinguishable symbols! Also, my light gray grid lines were too prominent when reproduced. When I complained, I was told that I was the only one who had submitted two versions.

Proofread graphs.

It is at least as important to proofread your graphs as it is to proofread your text. I find this principle one of the most difficult to control. The problem occurs when you submit figures for publication and are not given the chance to review the document before it comes out. Another time I submitted a column to a newsletter. They changed the shape of the figures, distorting the point I was trying to make. I wish that I could offer a solution to these problems. My only advice is to proofread before you submit, have someone else check, and request that you review your work before publication.

The chart or graph must be consistent with the text.

It is not uncommon to see text describing a graph that is not consistent with the figure. This often occurs when the writer changes one without thinking to change the other. Once again, careful proofreading is required.

Fig. 6.20 World Population Data

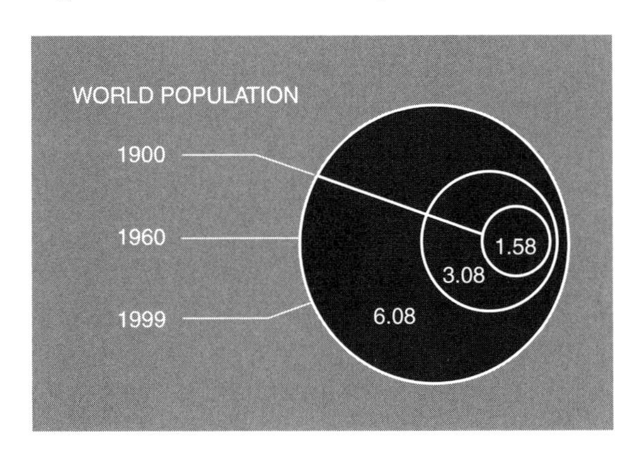

Figure 6.20 (Wurman, 1999, Rao chapter) has three circles and three labels, but two of the labels point to the same circle. In this case the reader can figure out the correction since the subject matter is well known and easily understood, but that is not always the case.

That is not the only problem with Figure 6.20. The diameters are proportional to the population data, but we visualize areas. A population of 6 billion is four times as large as a population of 1.5 billion, but the area is 16 times as large. Playfair (1801) used circles to represent population over 200 years ago,[3] but he did so by making the areas, not the diameters, proportional to the population.

[3] See Figure 2.18.

Fig. 6.21 Internet Users Data

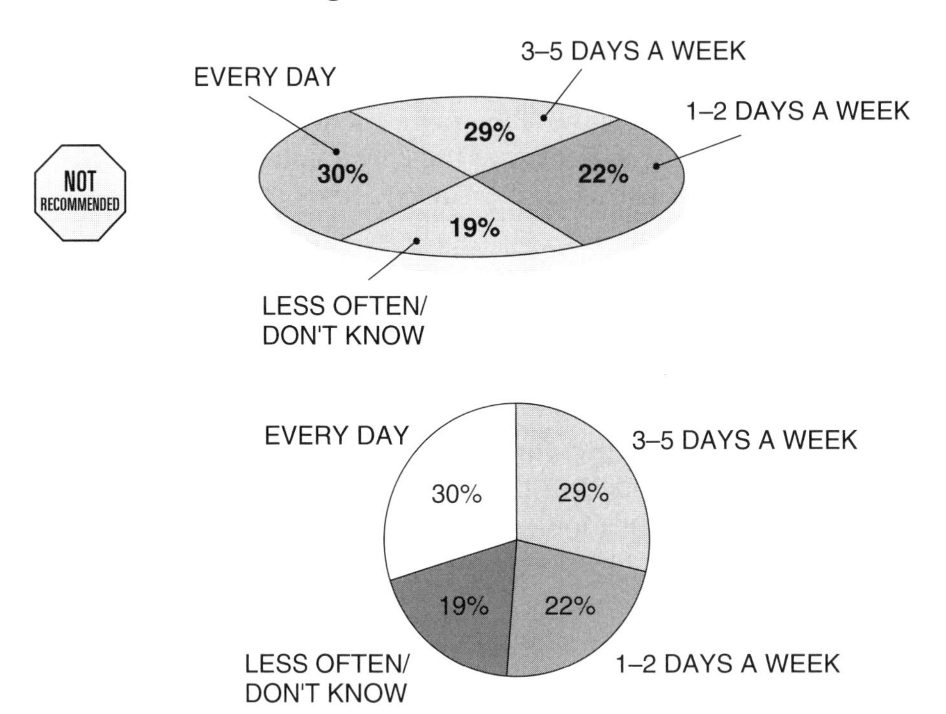

6.3 CLEAR UNDERSTANDING

Draw the data to scale.

Graphs are pictorial representations of numerical quantities. It therefore seems reasonable to expect that the visual impression we get when looking at a graph is proportional to the numbers that the graph represents. Unfortunately, this is not always the case. Look at the labels on the wedges of Figure 6.21 (Wurman, 1999, Green and Donovan chapter),[4] which shows how often Internet users go online. The wedge labeled 22% and the one labeled 30% appear to be approximately the same size and the 19% wedge appears more than two-thirds the size of the 29% wedge. The bottom chart shows the wedges drawn accurately. I innocently assumed that the common use of computers assured that most graphs were drawn to scale, but the frequent occurrence of graphs like this one prove me wrong. Therefore, check your graph carefully to make sure that the visual impression makes sense in terms of the data.

[4] At the time of this writing, this figure is available in original color at *http://www.understandingusa.com/chaptercc=13&cs=292.html*

Fig. 6.22 Police Officer Data

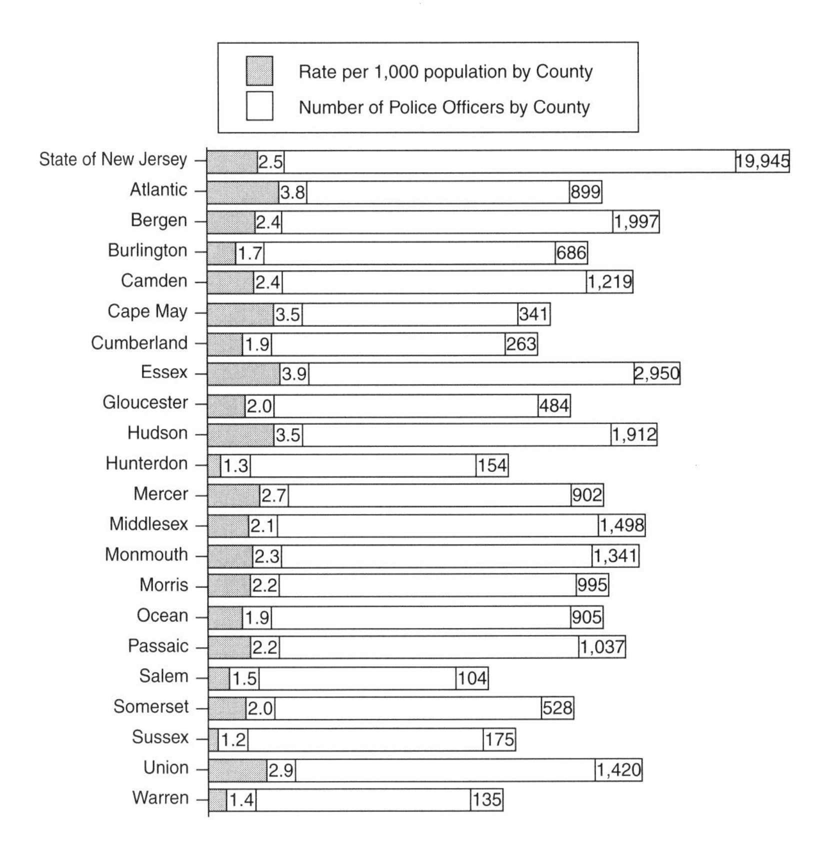
NUMBER OF MUNICIPAL POLICE OFFICERS
AND
RATE PER 1,000 POPULATION BY COUNTY 1997
Rate per 1,000 population by County
Number of Police Officers by County
State of New Jersey 2.5 19,945
Atlantic 3.8 899
Bergen 2.4 1,997
Burlington 1.7 686
Camden 2.4 1,219
Cape May 3.5 341
Cumberland 1.9 263
Essex 3.9 2,950
Gloucester 2.0 484
Hudson 3.5 1,912
Hunterdon 1.3 154
Mercer 2.7 902
Middlesex 2.1 1,498
Monmouth 2.3 1,341
Morris 2.2 995
Ocean 1.9 905
Passaic 2.2 1,037
Salem 1.5 104
Somerset 2.0 528
Sussex 1.2 175
Union 2.9 1,420
Warren 1.4 135

NOT
RECOMMENDED

Figure 6.22 shows both the number of police officers by county in New Jersey and the rate per 1000 population (State of NJ Uniform Crime Reporting Unit, 1997). The rate is not quite to scale; note that the length of 1.2 for Sussex County is less than half of the length of 2.4 for Bergen or Camden County. The number of police officers isn't even close. I didn't need my trusted ruler to discover that the length of the line for Salem County, which should be about a tenth of the length of the line for Passaic County, is much too long. Graphs should help to visualize comparisons; they cannot do so when the pictures are not proportional to the numbers they represent.

Fig. 6.23 Police Officer Data to Scale

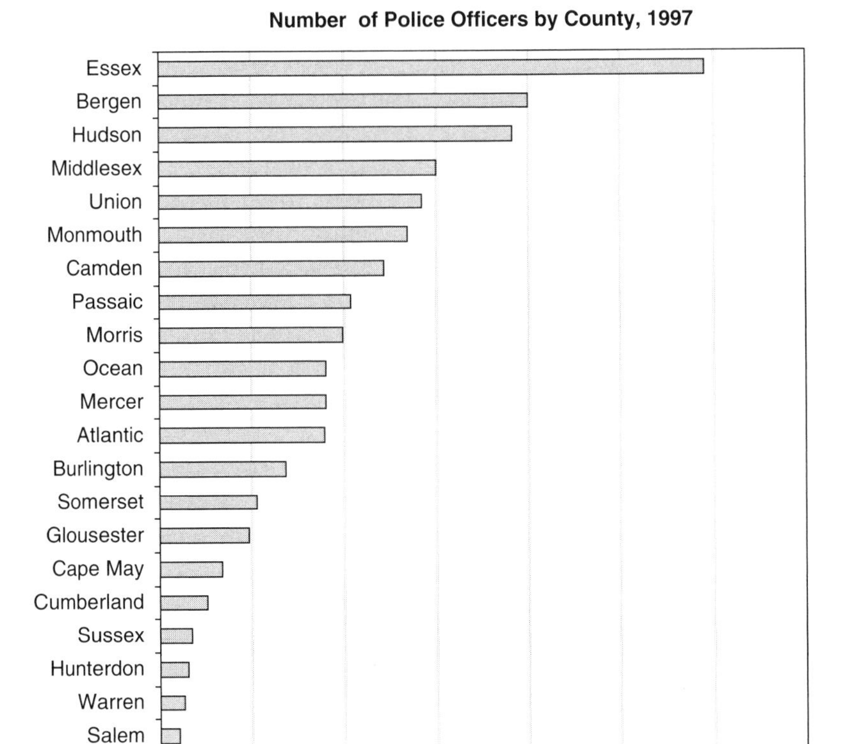

Here we see the number of police officers by county using the same type of graph as in their report. Figure 6.23 aids comparisons and helps to visualize the numbers. I prefer the rate

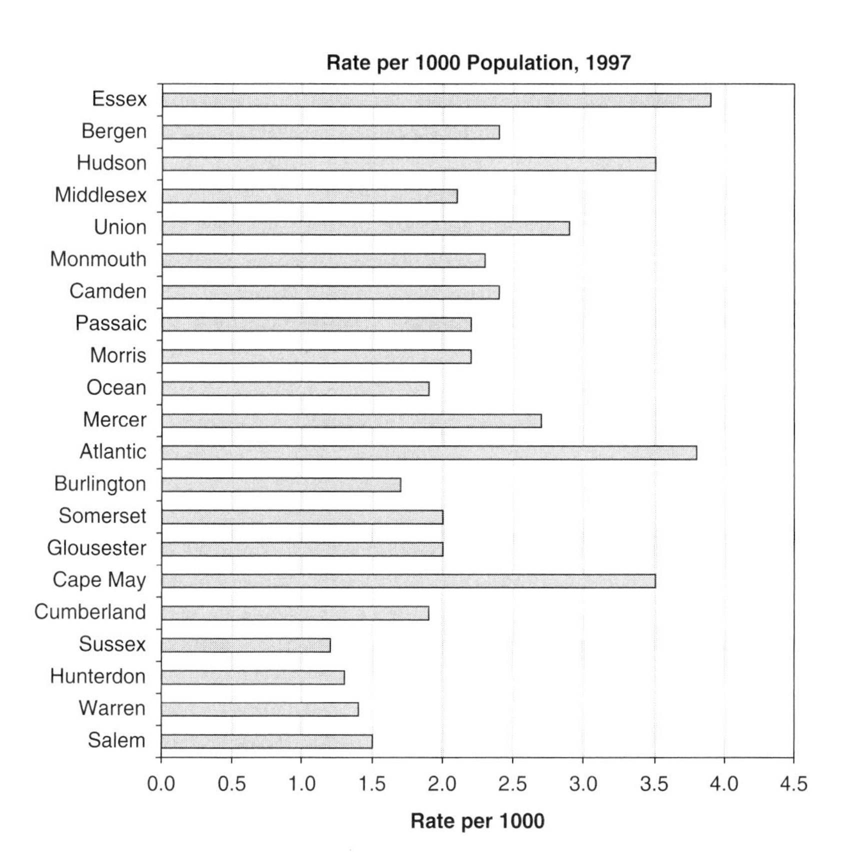

per 1000 population on a separate chart with a scale appropriate for its numbers, as shown. The counties for the rate plot are ordered by their number of officers, for consistency.

Fig. 6.24 Earnings per Share Data

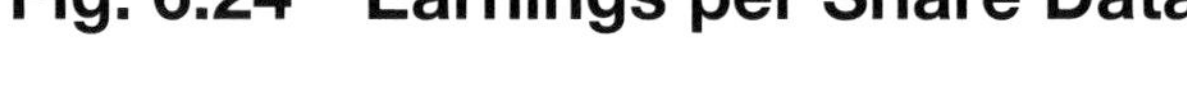

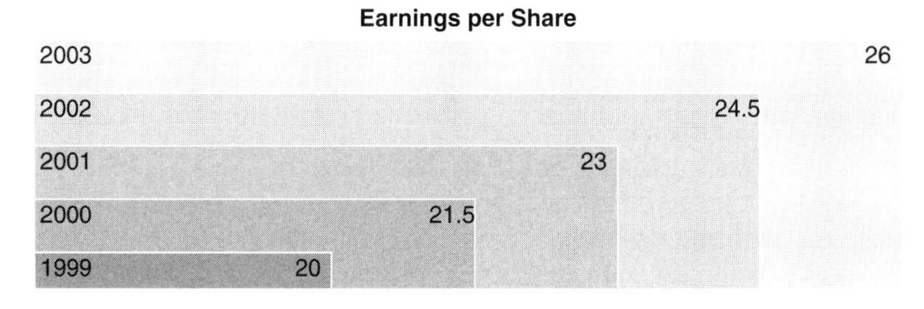

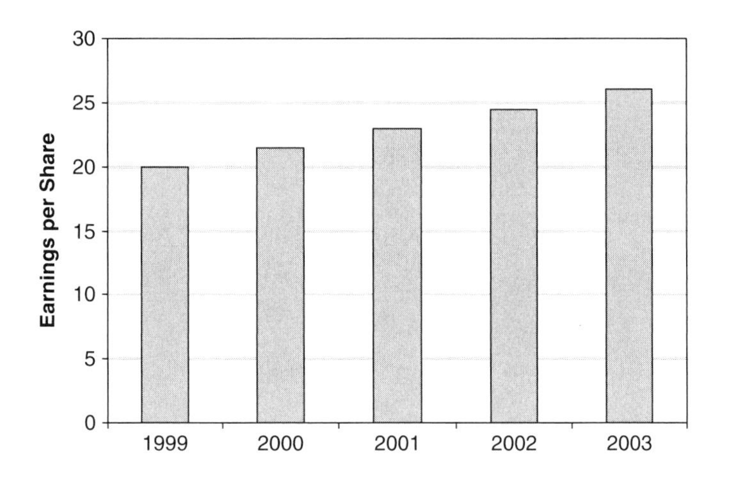

Do not show changes in one dimension by area or volume.

The top graph in Figure 6.24 shows earnings per share for a company from 1999 to 2003 as rectangles, whose length presumably corresponds to an imaginary x axis measured in dollars. Two-dimensional figures, however, are misleading when only one dimension changes, since they will cause the reader to overestimate the relative size of the larger numbers. For example, the largest rectangle in Figure 6.24 represents 26, a number 1.3 times as large as the value of the smallest rectangle, 20. But the area of the largest rectangle is over 20 times the area of the smallest. Thus, the *lie factor* (Tufte, 2001, p. 57), the size of the effect in the graphic divided by the size of the effect in the data, is over 16. An honest graph has a lie factor of 1. A second problem is that the baseline is not zero even though we are judging areas. That contributes to making the increase in earnings per share appear much larger than is actually the case. Compare the visual impression you get from the top plot with the honest one at the bottom.

Fig. 6.25 America Online Market Value

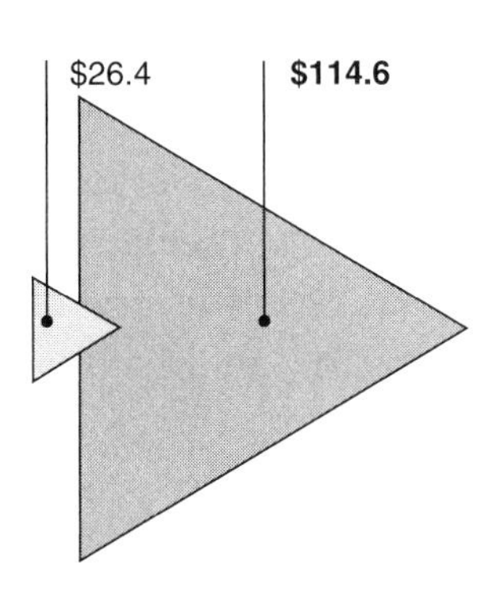

+**334.1%**

America Online

$26.4 **$114.6**

The top of Figure 6.25 is the America Online segment of a figure that compares the market value of Forbes 500 companies from 1998 to 1999 in billions of dollars, to demonstrate that Internet companies rose in value much more than traditional companies rose during that time period[5] (Wurman, 1999, Green and Donovan chapter). The left triangle shows the market value for 1998, and the right triangle, for 1999. However, it is not clear what corresponds to the $26.4 billion and the $114.6 billion: the area of the triangles, the sides of the triangles, the altitudes of the triangles, or what? Careful measurement shows that the figure is proportional to the numbers if the altitudes of the triangles represent the company values. Most people visualize the areas, thereby overestimating the increase. In the bottom figure, the areas are proportional to the dollar values.

[5] At the time of this writing, the full figure in color can be seen at *http://www.understandingusa.com/chaptercc=13&cs=286.html.*

Fig. 6.26 Annual Report Data

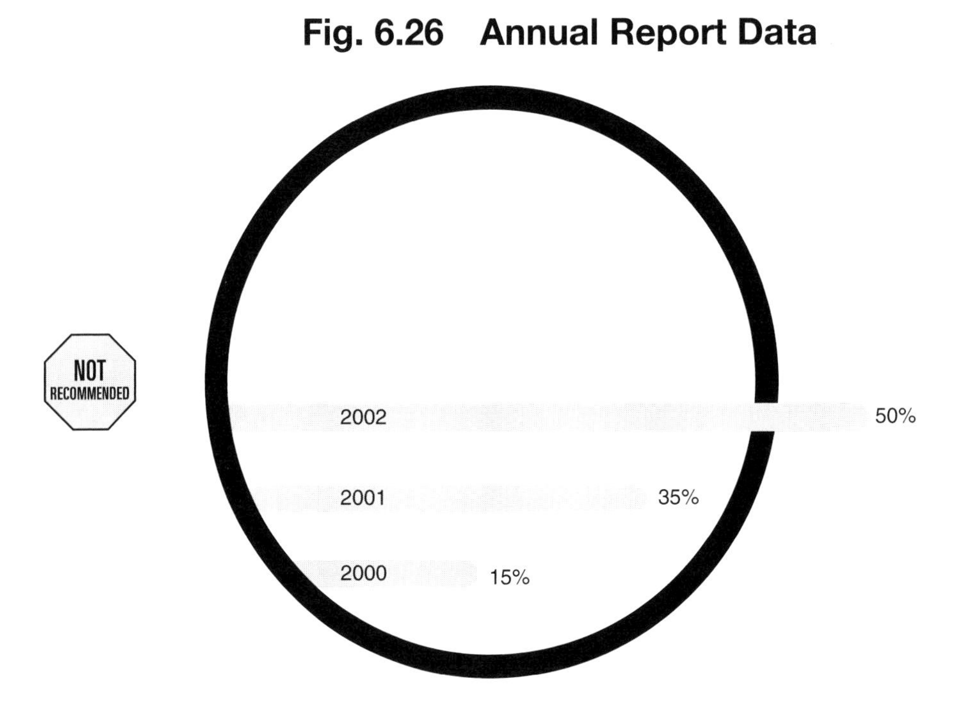

Use a common baseline wherever possible.

Figure 6.26, suggested by similar charts seen in corporate annual reports, starts the bars on the left side of the circle. It is not clear how to determine the length of the bars: Do we read the 2000 bar from the top closer to the 2001 bar or from the bottom? Perhaps it should extend to the baseline of the 2002 bar. I measured all three of these methods and discovered that the data are not drawn to scale. Using different baselines distorts the comparison of the bars, and not drawing the bars to scale distorts the data further.

Fig. 6.27 Immigration Data

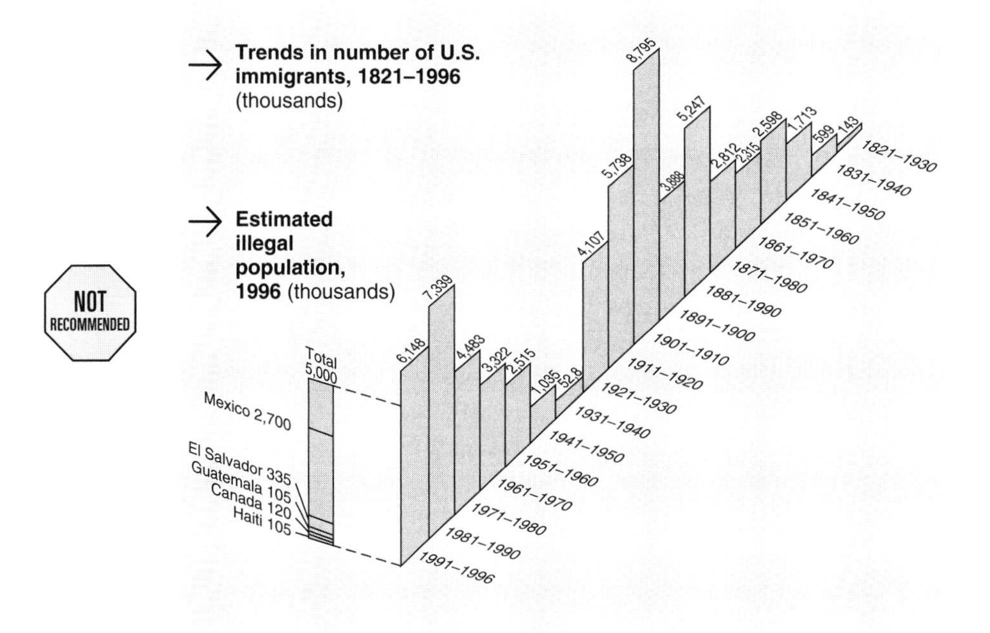
NOT
RECOMMENDED
Trends in number of U.S. immigrants, 1821–1996 (thousands)
Estimated illegal population, 1996 (thousands)
Total 5,000
Mexico 2,700
El Salvador 335
Guatemala 105
Canada 120
Haiti 105
6,148
7,339
4,483
3,322
2,515
1,035
52.8
4,107
5,738
8,795
3,888
5,247
2,812
2,315
2,598
1,713
599
143
1821–1930
1831–1940
1841–1950
1851–1960
1861–1970
1871–1980
1881–1990
1891–1900
1901–1910
1911–1920
1921–1930
1931–1940
1941–1950
1951–1960
1961–1970
1971–1980
1981–1990
1991–1996

The bars in Figure 6.27 (Wurman, 1999, Lenk and Kahn chapter) lack the common horizontal baseline that facilitates judgments about length. Also, the convention in charts is to read time from left to right. Readers assume this to be the case and sometimes miss the labels when the time direction is reversed.

Fig. 6.28 Immigration Data with Horizontal Baseline

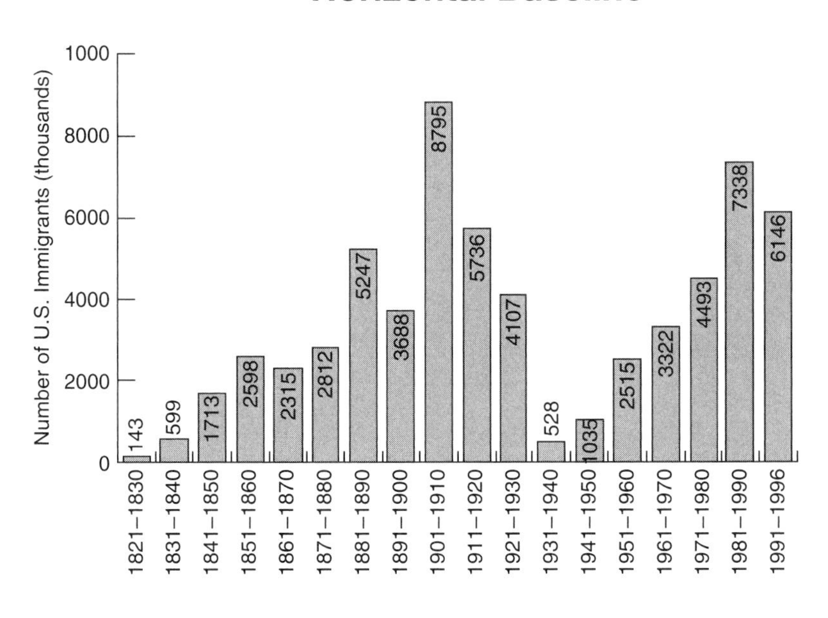

Figure 6.28 (Tracey, 2004), drawn using Illustrator, shows the same data drawn in a standard two-dimensional vertical bar chart for comparison. Most readers will find it easier to understand the trend. A dot plot would also show this information clearly.

Fig. 6.29 Survey Results Data

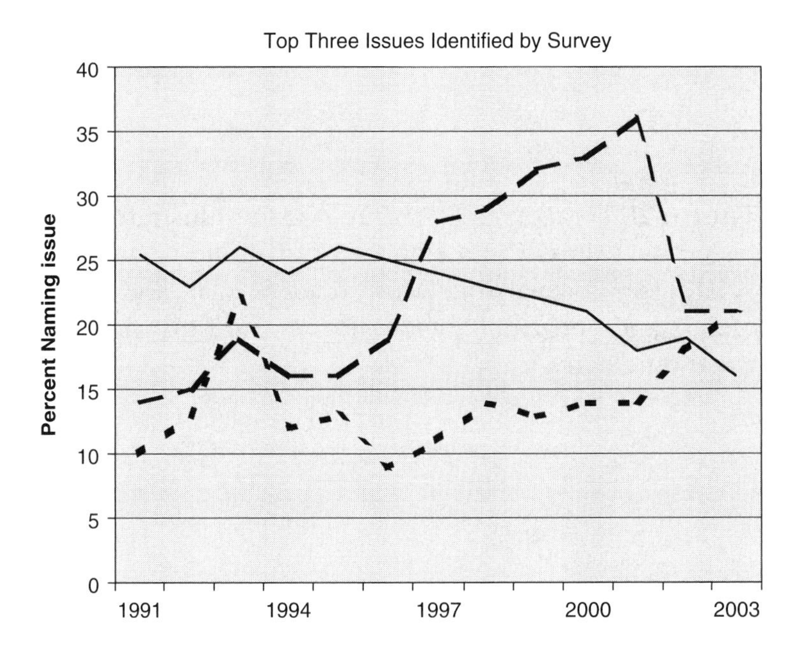

Label data sets directly when it doesn't clutter the graph.

The percentage of survey respondents naming each of three issues as the main problem is shown in Figure 6.29. There are two common ways to label the three issues: within the scale-line rectangle close to the line to which the label refers, or in a key or legend. Labeling within the rectangle avoids having to look back and forth to identify the data, while legends and keys avoid clutter. In general, it is preferable to label the lines directly but to use legends when direct labeling crowds the figure. Unfortunately, the creators of Figure 6.29 forgot to label the lines either way. Other problems with labels include reversing the labels, unclear labels, and labels in different units than the figures.

Fig. 6.30 Health Insurance Data

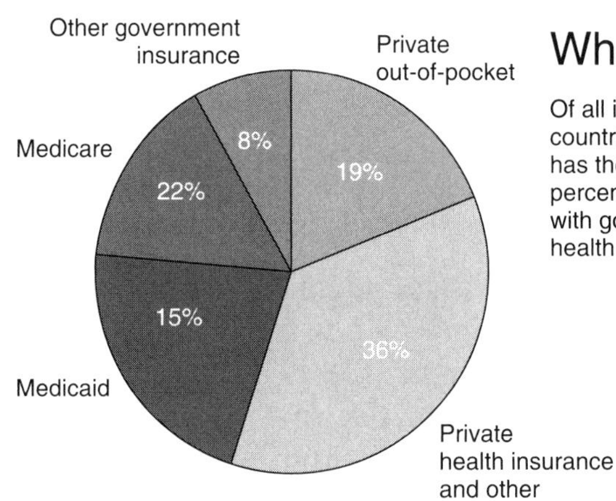

Who pays?

Of all industrialized countries, the U.S. has the lowest percentage of population with government-assured health insurance.

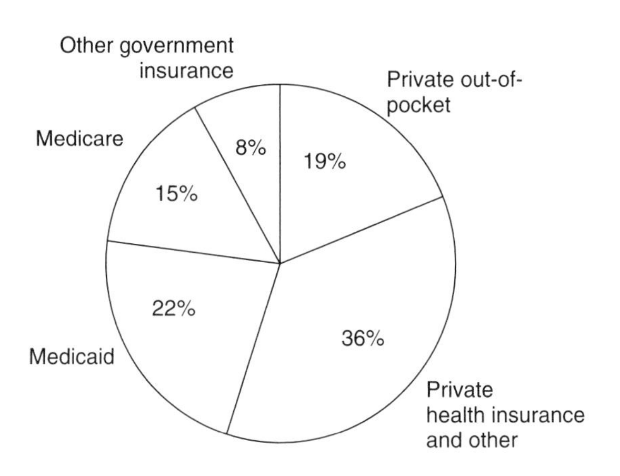

At first it appears that the top plot in Figure 6.30 is not drawn to scale because the wedge labeled 15% appears larger than the one labeled 19% and much larger than the one labeled 22% (Wurman, 1999, Bierut chapter).[6] But redrawing it shows that the wedges appear to be correct. The problem is that two labels are reversed; the wedge labeled 15% should be 22% and the one labeled 22% should be 15%. I don't know whether the category labels (Medicare and Medicaid) are reversed.

[6] At the time of this writing, this figure is available in original color at *http://www.understandingusa.com/chaptercc=7&cs=150.html*.

Don't require the reader to make calculations that a computer can make more easily.

Examples of this problem appear several times in this book. Figure 2.14 illustrated Playfair's data on exports and imports. We saw that if the reader is interested in the balance of trade, it is better to plot the balance of trade directly than to require the reader to make subtractions. Again, in Figure 8.3, you will see an example where the *before* figure requires the reader to subtract mentally.

There are many situations where *before* and *after* data are presented. Some examples include weight before and after a diet, SAT scores before and after tutoring, and medical measurements before and after medication. In each of these cases we are interested in the change caused by the intervention. Readers will understand the data better if we plot the improvement or deterioration directly rather than require them to calculate it.

Therefore, another way to state this principle follows:

Plot the variable of interest. If interested in improvement, plot improvement rather than *before* and *after*.

Fig. 6.31 Source of Funding Data

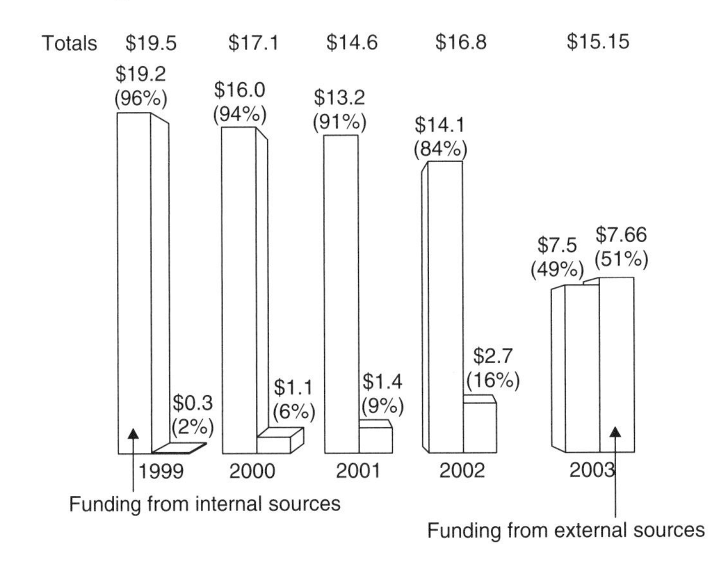

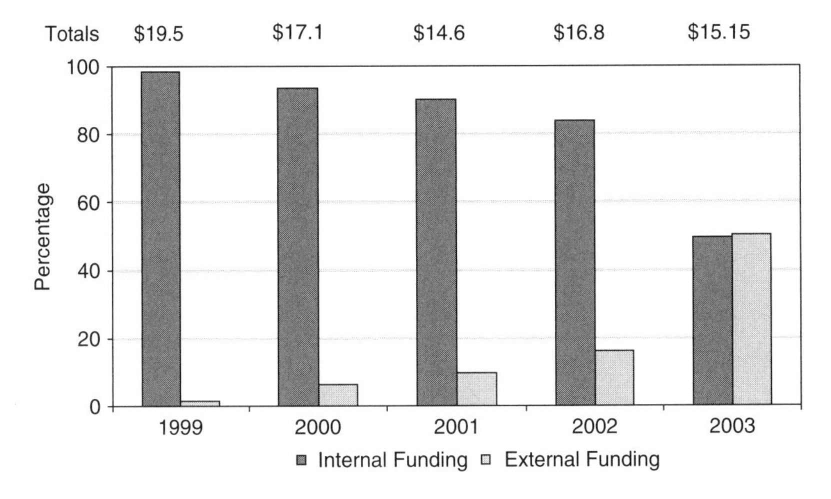

Strive for clarity.

At first glance the top chart in Figure 6.31 appears to be a grouped three-dimensional bar chart. It shows the sources of funding for a nonprofit organization. The left bar in each group is the funding provided by internal sources, and the right bar represents funding from outside sources. The reader at first is confused that 14.1 in 2002 is shorter than 13.2 in 2001. A closer inspection shows that the size of the bars refers to the percentage of internal funding, not the actual number of dollars. To make this clear, the percentage should be in bold type and the actual number of dollars, which is of secondary importance for the point the author is emphasizing, should be in parentheses.

Fig. 6.32 Population of Cities

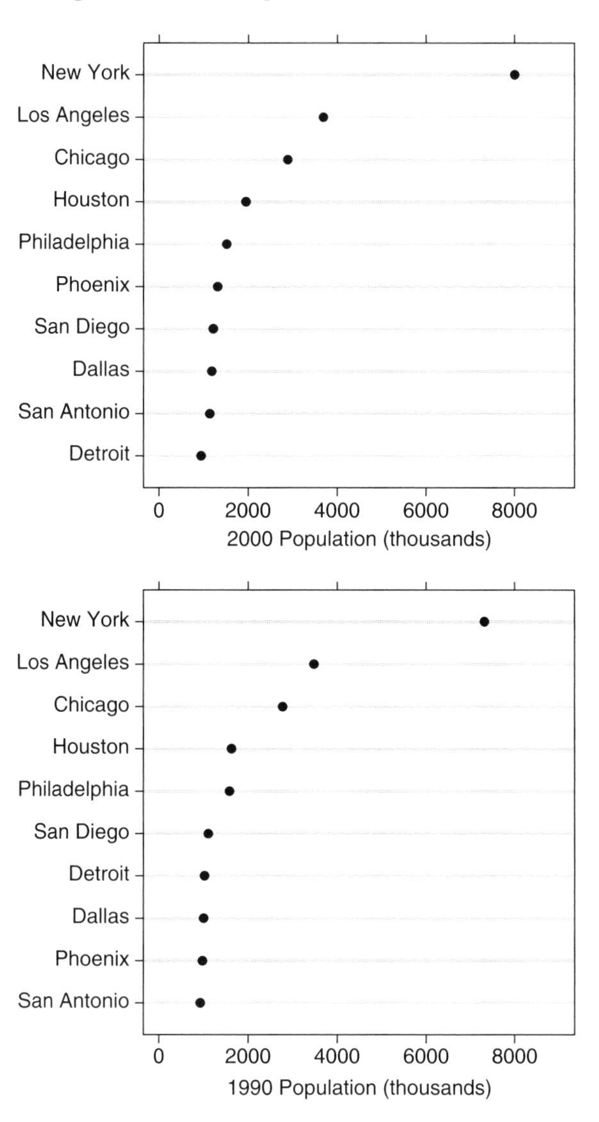

Groups of charts need consistency in order, color, and other graphical elements.

The top plot of Figure 6.32 shows the 10 largest U.S. cities by population in 2000; the bottom figure, in 1990. Although the same 10 cities appear in both figures, the size order has changed in some cases and each chart is ordered by size. For example, Phoenix was ninth in 1990 and sixth in 2000. Some readers might assume that the order is the same on both charts—and be misled.

I have seen groups of charts in which the color is reversed from chart to chart. If we are comparing data from A and B and red represents A on the first chart while blue represents B, using blue for A and red for B in the second chart confuses the reader. Consistent scales are discussed in Chapter 7.

Fig. 6.33 Population of Cities: Conflicting Principles

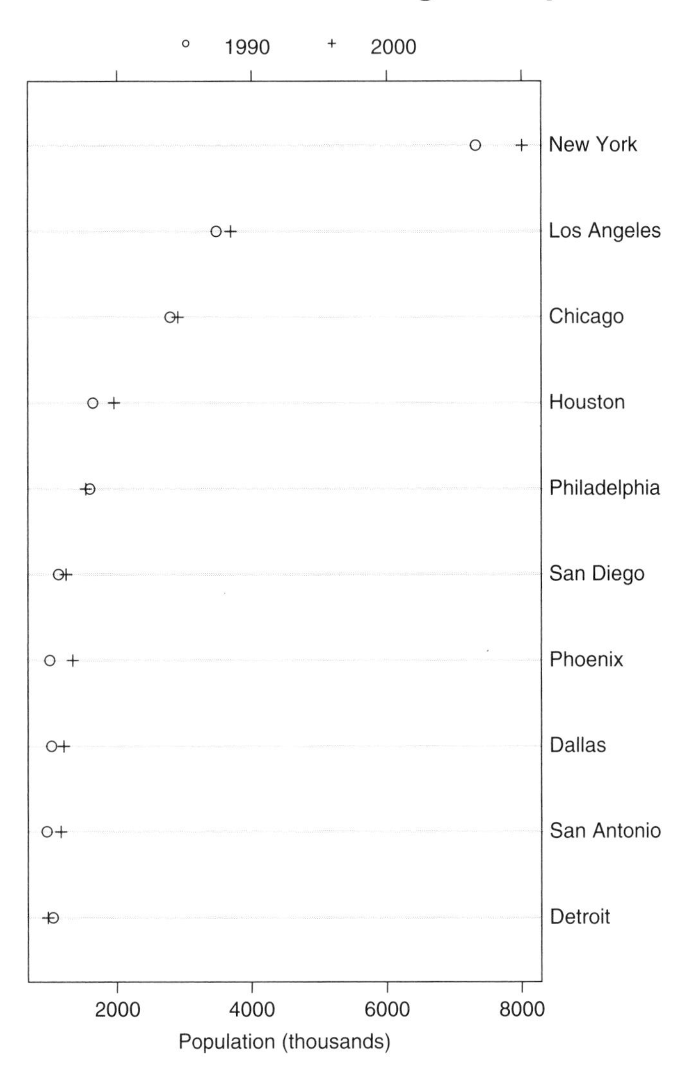

Choose the principle least likely to mislead if more than one applies and they conflict with one another.

There will be times when more than one of the principles in this book apply to the figure you are trying to create and they conflict with one another. For example, we have shown that ordering data by size is often more informative than ordering in other ways. We have also emphasized the need for consistency when making comparisons. If we want to compare the 10 largest U.S. cities in 1990 and 2000, which of these principles do we follow? Do we order each separately by size, or do we use the same order for both? To decide, consider which is more likely to confuse or mislead the audience. That is the one to avoid. In this example it is more important to be consistent than to order each by size. Another option is to look for another graph form that avoids the conflict, such as Figure 6.33, ordered by the 2000 population.

6.4 GENERAL STRATEGY

A number of principles we already discussed, such as making the data stand out and drawing the data to scale, offer general strategies for creating effective graphs. Other strategies follow:

- ***A large amount of quantitative information can be packed into a small region.*** The carbon dioxide data, the stock market data, and other examples that we have seen contain hundreds or even thousands of data points. These are uncluttered graphs. Many of the most cluttered graphs are attempts to disguise limited amounts of data.
- ***Graphing data should be an iterative, experimental process.*** Frequently, a first attempt at a graph suggests other ways of looking at data and other questions about the data that a new presentation might answer.
- ***Graph data two or more times when needed.*** Different presentations of data emphasize different aspects of the data. It is often useful to see more than one. In trellis displays it is often helpful to condition on the variables in different orders.

- ***Many useful displays require careful, detailed study.***
  One often hears that the main message of a good graph jumps at you immediately. Our definition of an effective graph claims that the quantitative information can be absorbed more quickly than with other presentations. That does not imply that it is absorbed quickly. A graph with a large amount of information may require detailed study.

SUMMARY

The reader can see and understand the data and other graphical elements of an effective graph. Principles are provided to ensure that this is the case, and examples of poor graphs are shown.

7 Scales

Scales (the rulers along which we graph data) have a profound effect on our interpretation of graphs. In this chapter I provide answers to the following questions on scales:

- What is the optimum aspect ratio? The *aspect ratio* is the ratio of the height of the data rectangle to its width.
- Must zero be included? This is one of the most controversial topics in constructing and evaluating graphs.
- When do logarithmic scales improve clarity? Readers of graphs are often puzzled by logarithmic axes.
- What are breaks in scales, and how should they be used? Breaks in scales often go unnoticed or are misused.

Creating More Effective Graphs, by Naomi B. Robbins
ISBN 0-471-27402-X Copyright © 2005 John Wiley & Sons, Inc.

- Are two scales better than one? How can we distinguish between informative and deceptive double axes? Examples are given of both deceptive and legitimate uses of double-y axes.
- Can a scale "hide" data? How can this be avoided? The variation of a variable can be hidden with the wrong choice of a scale.

The word *scale* has more than one meaning: It can refer either to the ruler by which we measure the data or to the number of units per inch. We use the former meaning.

Fig. 7.1 Sunspot Data: Aspect Ratio 1

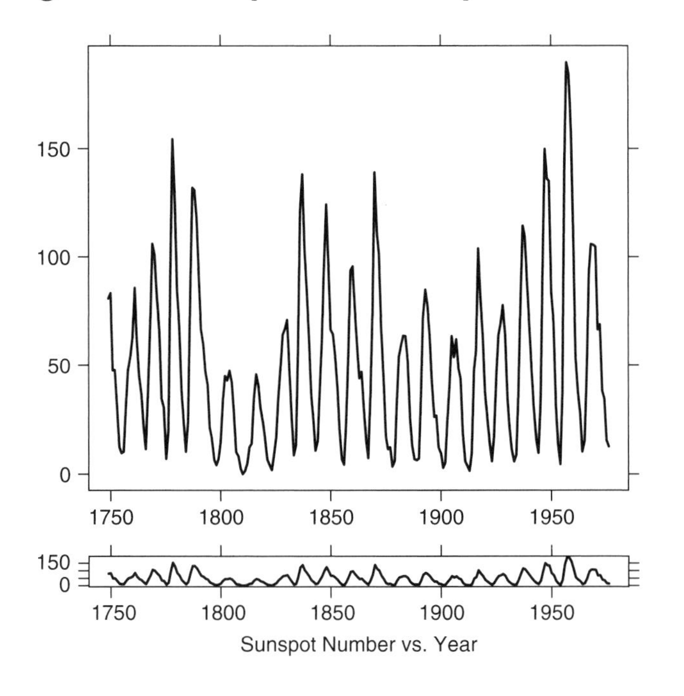

7.1 ASPECT RATIO: BANKING TO 45°

The choice of aspect ratio can greatly affect our perception of data: the two parts of Figure 7.1 show the same data, the top graph with an aspect ratio of 0.8, and the bottom graph, one of 0.055. They both chart the frequency of sunspots from 1749 to 1976. The first thing we notice in both figures is the cyclical nature of the data. These cycles have an average period of about 11 years. But there is another important feature of the data that is seen much more clearly in the lower panel. The rise of the cycles is much steeper than the fall. This property is most pronounced in the cycles with the highest peaks and is critical for solar physicists in developing theories that explain the origin of sunspots.

Fig. 7.2 Annual Report: Aspect Ratio 2

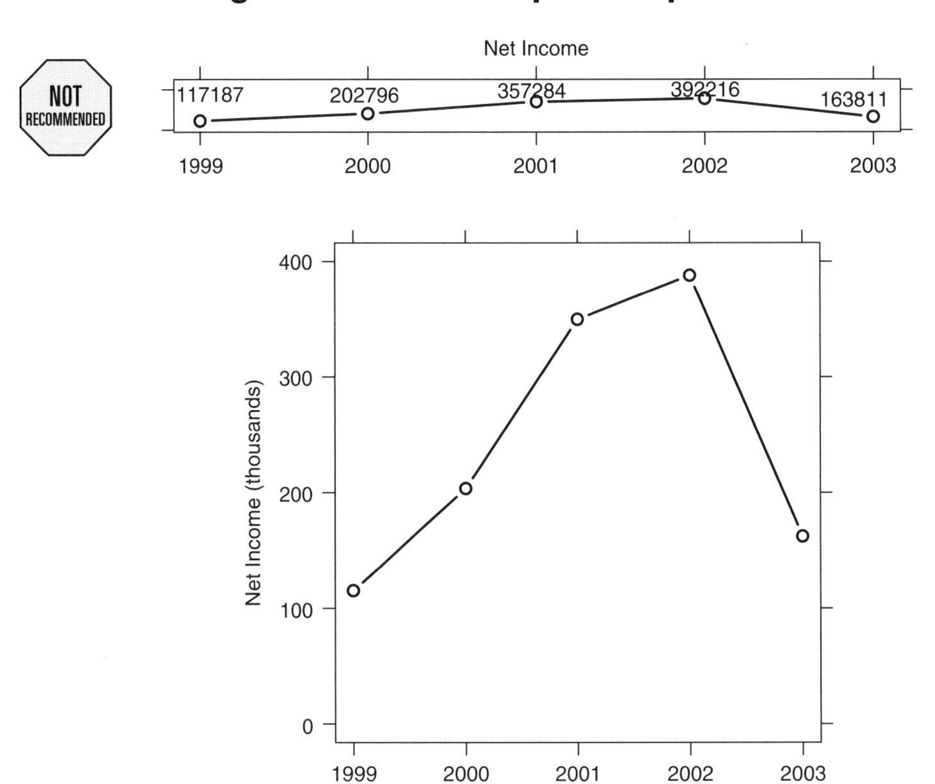

For almost a century contradictory opinions on the optimum aspect ratio appeared. The orientation of a line segment is its angle with the horizontal. These orientations change as the aspect ratio changes. The human eye can judge a 45° angle with considerable accuracy. Small angles such as 5° and 10° are much more difficult to judge. Designers can take advantage of this by changing the aspect ratio of a chart so that the orientations of line segments are centered at 45° and can be assessed more accurately (Cleveland et al., 1988). Called *banking to 45°*, this is how the aspect ratio of the bottom panel of Figure 7.1 was chosen. The trellis graphics of S-Plus and the lattice package of R allow you to determine the aspect ratio and bank to 45°. You can approximate these ideas by trial and error using everyday software.

Again, both the top and bottom parts of Figure 7.2 show the same data. The top figure is similar to those seen in corporate annual reports. The aspect ratio disguises the large drop in net income. Do you think that this was intentional?

7.2 SCALES: MUST ZERO BE INCLUDED?

Whether zero needs to be included in a scale is probably the most controversial topic in the presentation of graphs. Cleveland (1994) says:

> Do not insist that zero always be included on a scale showing magnitude.
>
> When the data are magnitudes, it is helpful to have zero included in the scale so we can see its value relative to the value of the data. But the need for zero is not so compelling that we should allow its inclusion to ruin the judgment of the variation in the data.

On the other hand, in a unit of their book, whose aim is "to expose sinister charlatans who deliberately use statistics to mislead people," Downing and Clark (1996) explain: The rule is that a graph of a change in a variable with time should always have a vertical scale that starts with zero. Otherwise, it is inherently misleading."

A supplemental magnified view of one area of a graph is the only proper use of a scale without zero, these authors note, since the reader has access to the "undistorted" graph. An excellent discussion of this topic appears in Cleveland (1994). The carbon dioxide trend line we saw in Figure 4.19 and repeat in the top plot of Figure 7.3 was actually used for testimony before the U.S. Senate on global warming. Note the increase in the slope or rate of change. Albert Gore, Jr., then a U.S. senator, used the curve to demonstrate increases in carbon dioxide. N. Douglas Pewitt, a witness for the Department of Energy, called the chart deceptive, saying: "It is a clever piece of chartology, in that it is intellectually accurate but can be subject to being read the wrong way."

Fig. 7.3 Carbon Dioxide Data

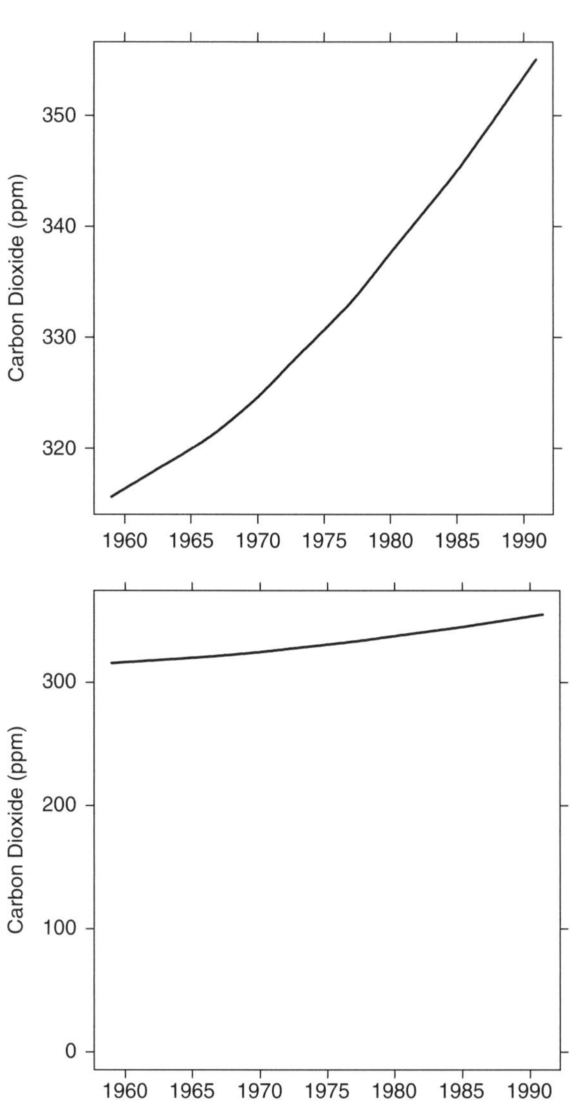

The bottom chart in Figure 7.3 shows the same data with a zero baseline. Recall that the *aspect ratio* is the ratio of the height to the width of the data rectangle and not of the scale-line rectangle. The aspect ratio of the bottom chart is too small to show the important fact that the rate of change is increasing. Legislators determining public policy need to know this. For this audience, I consider the chart with the zero origin to be the misleading one since it hides information the reader needs to know.

Fig. 7.4 Carbon Dioxide: Aspect Ratio Changed

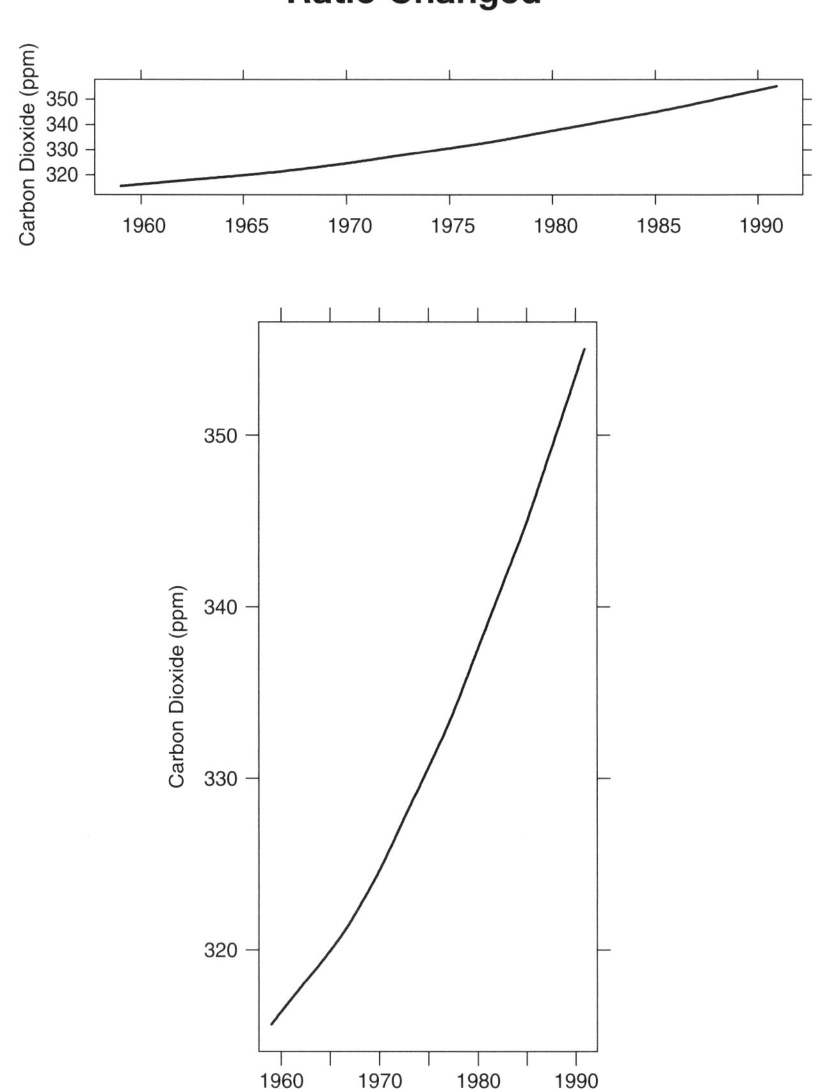

A zero baseline is not needed to hide the variation of the data or to exaggerate growth; changing the aspect ratio can have the same effect. The top chart in Figure 7.4 shows just the part of the zero baseline chart that contains data. The bottom chart has an aspect ratio of 2. They give a very different impression of the rate of growth to a reader who does not study the tick marks and labels.

Now imagine the two carbon dioxide charts of Figure 7.3, but this time imagine that the vertical axis represents a tax rate. This time a political candidate is showing the graph to the public, including many who are not used to reading graphs and cannot be expected to read axis labels carefully. The candidate is stating that the upper curve represents the increase in tax rate while the opposing party was in power, and the lower chart represents the increase in the tax rate while the candidate's party was in power. With this audience, the upper graph gives a misleading impression of a large increase in taxes.

Fig. 7.5 Annual Report Bar Chart: No Zero

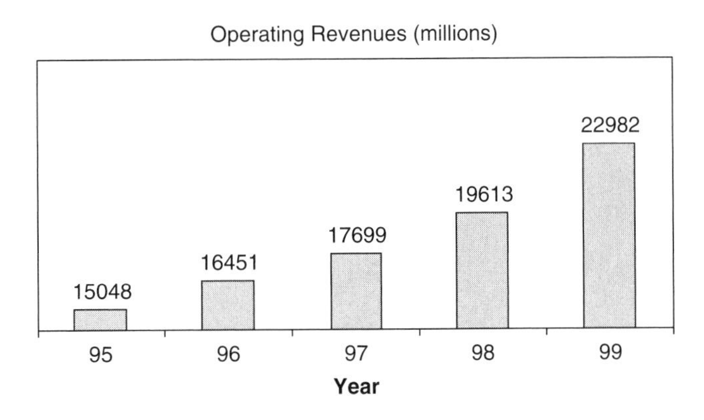

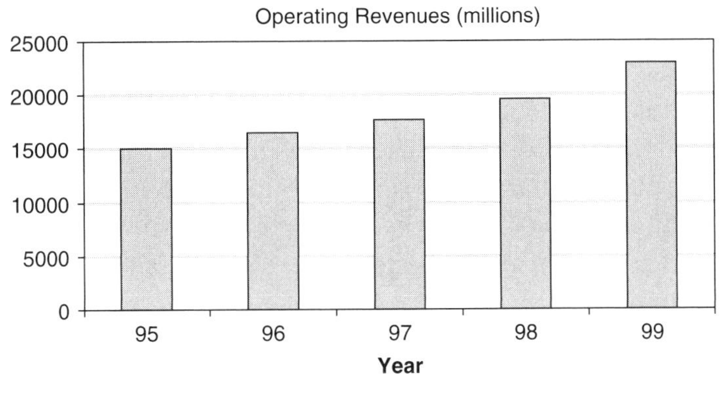

Whether or not zero must be included also depends on the type of graph being used. We judge the level of carbon dioxide in Figure 7.3 by the position along a common scale, and this position does not depend on where zero appears. On the other hand, when we read a bar chart, we see the entire area of the bar, and not including zero gives a distorted view of the area. This problem is common in annual reports of corporations. In fact, the top chart in Figure 7.5 is modeled after a real example from an annual report. There are no axis labels on the vertical axis and the baseline is not labeled. Around $16\frac{1}{2}$ billion appears almost twice the size of 15 billion, and slightly below 23 billion appears many times 15 billion. The visual impression is a lie.

My position is that any *bar chart* must include zero. However, the answer is not as clear for line charts or other charts for which we judge position along a common scale. We already saw that some audiences need to see the detailed variation in the data, whereas others may be misled by it. Whether or not to include zero depends on the situation and the audience.

In 1954, Darrell Huff wrote a delightful little book entitled *How to Lie with Statistics*. Although I highly recommend most of the book, the chapter on the *Gee Wiz Graph* has played a large role in spreading the myth that all graphs require a zero. In his all-day workshop on presenting data, Edward Tufte (2001) dramatically gets up on a table, opens a book, and unfolds a long narrow piece of paper until it falls to the floor. On the top of it is a small line graph; the rest is blank. Tufte asks the audience what purpose the 7 to 8 feet of blank paper serves, and says, "Darrell Huff was wrong."

Fig. 7.6 Logarithmic Scales for Percent Change

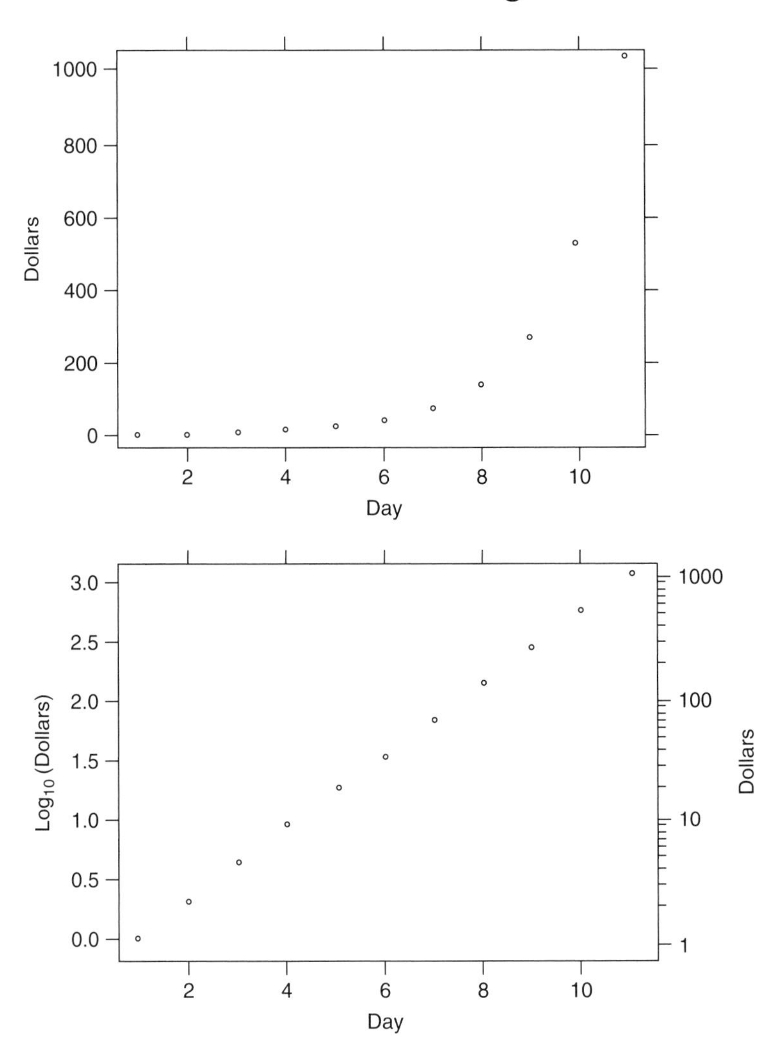

7.3 WHEN TO USE LOGARITHMIC SCALES

Use a logarithmic scale when it is important to understand percent change or multiplicative factors.

The top plot in Figure 7.6 shows the amount of money you would get if I gave you one dollar the first day, two dollars the second day, and I kept doubling the amount for 11 days. The bottom plot shows the same values with a logarithmic scale on the vertical axis.[1] Note that the constant percentage rate (doubling each value) produces a straight line on the logarithmic scale. On the linear scale absolute distances are constant: The vertical distance from the 200 tick mark to the 400 tick mark on the vertical scale of the top plot is the same as the vertical distance from the 400 tick mark to the 600 tick mark. On the logarithmic scale percentage differences are constant: The vertical distance from the 1 tick mark to the 10 tick mark on the right vertical axis of the bottom plot is the same as that from the 10 tick mark to the 100 tick mark. The linear scale is preferable if you want to know how many more dollars you received one day than another. The logarithmic scale is preferable if you are interested in how many times more money you received one day than the other.

[1] A graph with one logarithmic scale is called a semilogarithmic or semi-log graph.

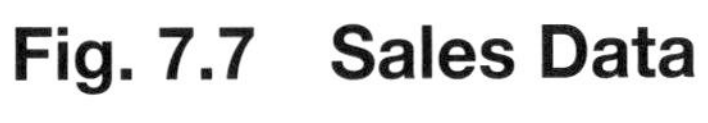
Fig. 7.7 Sales Data

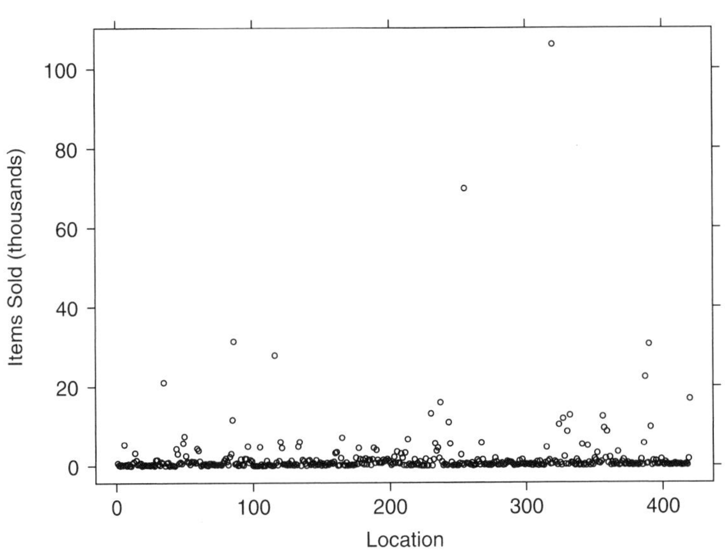
Items Sold (thousands)
0
20
40
60
80
100
0
100
200
300
400
Location

Showing data on a logarithmic scale can cure skewness toward large values.

We've already seen how a logarithmic scale cured skewness toward the large areas of Texas and Alaska. Here we show a business example. The data are real, but labels have been changed to preserve confidentiality.

Figure 7.7 shows the number of items sold in each of 420 locations (numbered 1 through 420). We notice immediately that a few locations sold orders of magnitude more than the others. Often, outliers similar to these are errors, but knowledge of the product and locations suggests that these large numbers are in fact reasonable.

This figure is not very informative since most of the data appear to be zero; the scale is determined by the large values, so there is no variation for the bulk of the data.

Fig. 7.8 Sales Data: Logarithmic Scale

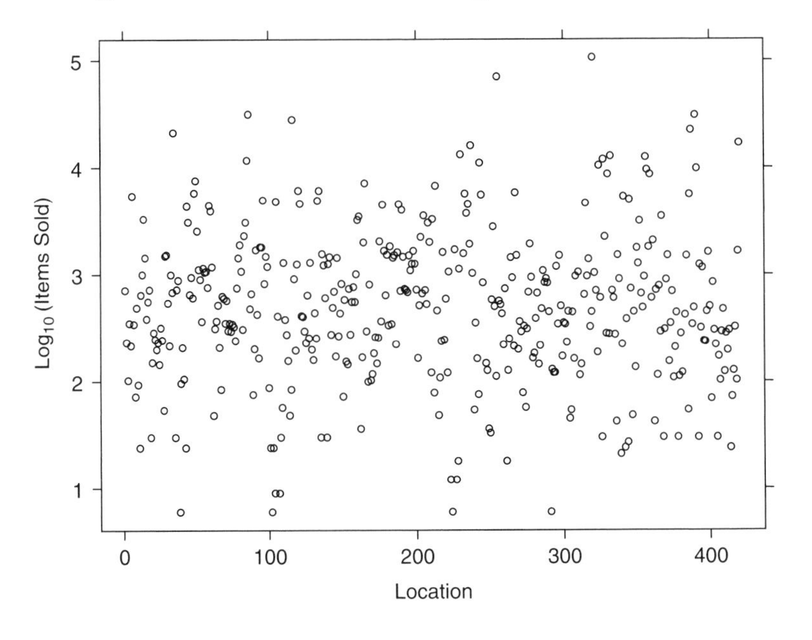

In Figure 7.8 we show the same data on a logarithmic scale with base 10. We get a better idea of the distribution of the data. The data appear most dense between 2 and 3. Since $\log_{10}(100) = 2$ and $\log_{10}(1000) = 3$, many locations sold between 100 and 1000 items. But just as an alphabetical ordering is often not the most informative, ordering by location number here can be improved upon.

Fig. 7.9 Sales Data: Ordered Data with Logarithmic Scale

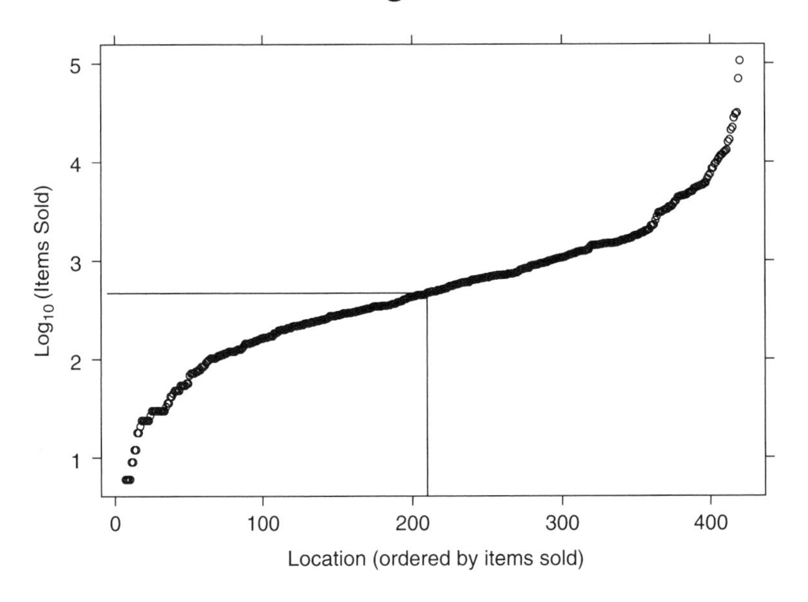

In Figure 7.9 we show the data ordered by the number of sales. This ordering shows the distribution of sales more clearly. It is easy to determine the median number of items sold. We know that there are 420 locations, so we draw a vertical line at 210.5 on the horizontal axis and then a horizontal line from the point where the vertical line intersects with the data to the vertical axis. Since the horizontal line reaches the vertical axis at about 2.67, and $10^{2.67}$ is around 470, roughly half of the locations had fewer than 470 sales and the other half had more sales. We can find other percentiles in a similar fashion.

Fig. 7.10 Police Data: Logarithmic Scale with Bars

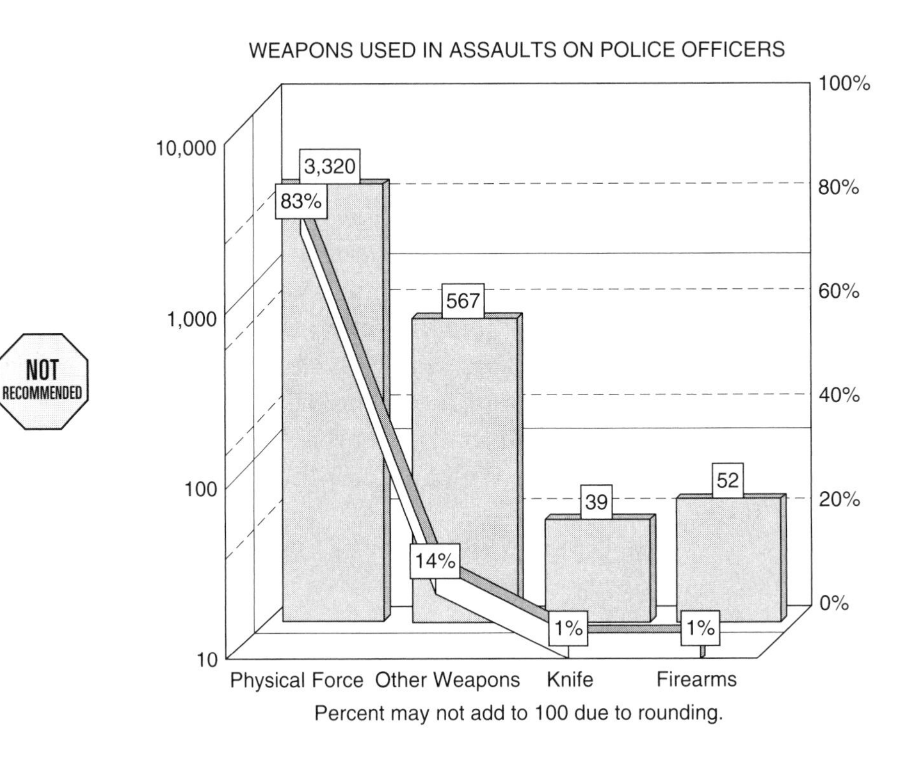

I have been saying that using logarithms can cure skewness toward large values. Figure 7.10 is an example that tries to follow this principle, but it doesn't work (State of NJ Uniform Crime Reporting Unit, 1997). The reader unconsciously compares areas when viewing bar graphs. It is confusing to see a bar labeled 567 that appears to be about two-thirds the height of a bar labeled 3320. Not all readers realize that they should read the bars from the left axis and the curve with percentages from the right. It is confusing that the absolute number is on a logarithmic scale and the percentage is not, so that visually the percentages do not match the bars. Many readers of this material may not be familiar with or may not remember logarithms.

Fig. 7.11 Police Data: Scale Break

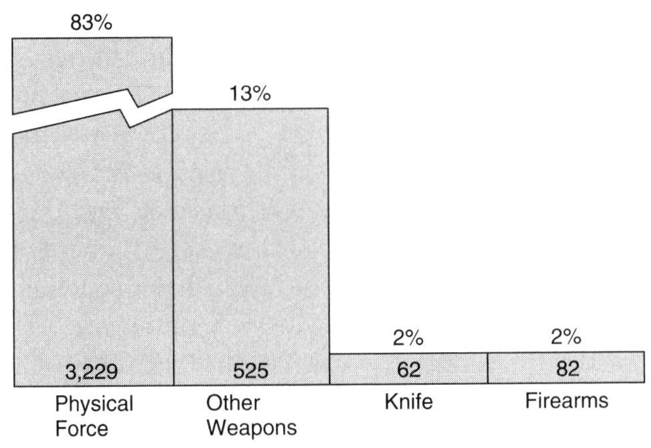

These police reports are published annually. Figure 7.10 reports 1997 data; Figure 7.11 reports 1994 data. We see that they used a scale break in earlier years. We get a feeling for the size of the data from the data labels and not the chart, so the chart serves no purpose. A table suffices for such a small data set. If a chart is desired, it should be understandable to the audience and show the comparisons of interest.

Fig. 7.12 Police Data: Simple View

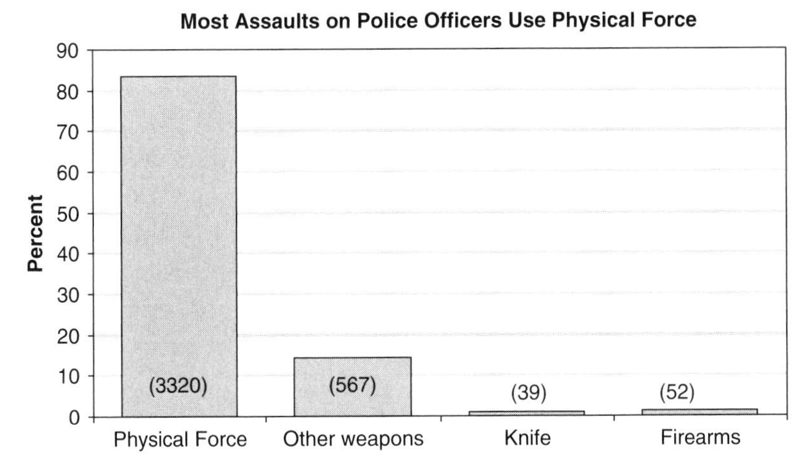

The bars represent the percentage of assaults in each category, and the numbers in parentheses represent the actual number of assaults.

Figure 7.12 emphasizes the fact that most assaults use physical force using hands, fists, or feet rather than other weapons, knives, or firearms. Both the relative size of the bars and the title emphasize this point. A table or even a sentence could communicate this small data set, but if a chart is desired, there is no need for scale breaks and logarithms. It is true that it is difficult to see the 1% bars, but that is exactly what these data are telling us. Emphasizing them is distorting the data.

Fig. 7.13 Sales Data: Full Scale Break

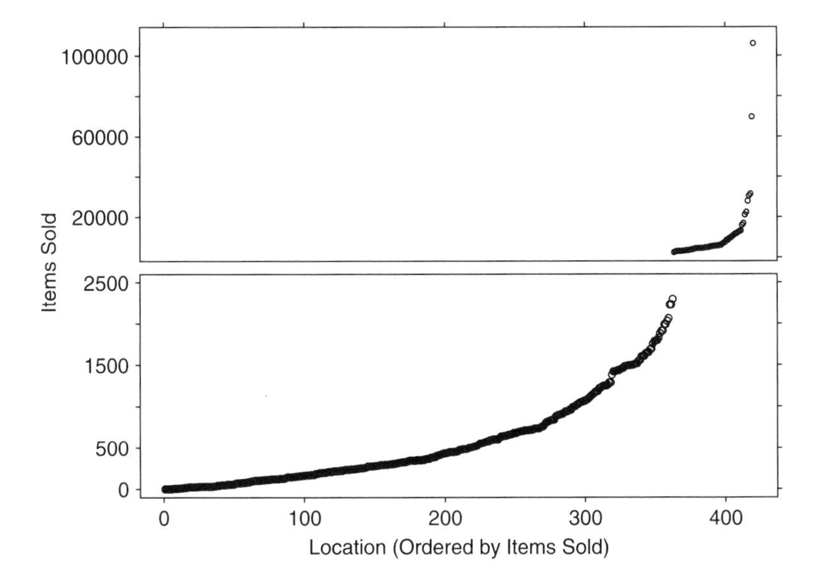

7.4 SCALE BREAKS

Use a scale break only when necessary. If a break cannot be avoided, use a full scale break. Taking logs can cure the need for a break.

We often see scale breaks indicated by two short parallel lines breaking an axis. These are often easy to miss. Figure 7.13 shows the sales data with a full scale break. The bottom panel shows the number of items sold for the bulk of the locations, while the top panel uses a different scale to show those locations with a large number of sales. We saw on previous pages that taking logarithms eliminated the need for a scale break. However, these figures might be more appropriate for an audience that is not familiar with logarithms.

Fig. 7.14 Sales Data: Avoiding a Scale Break

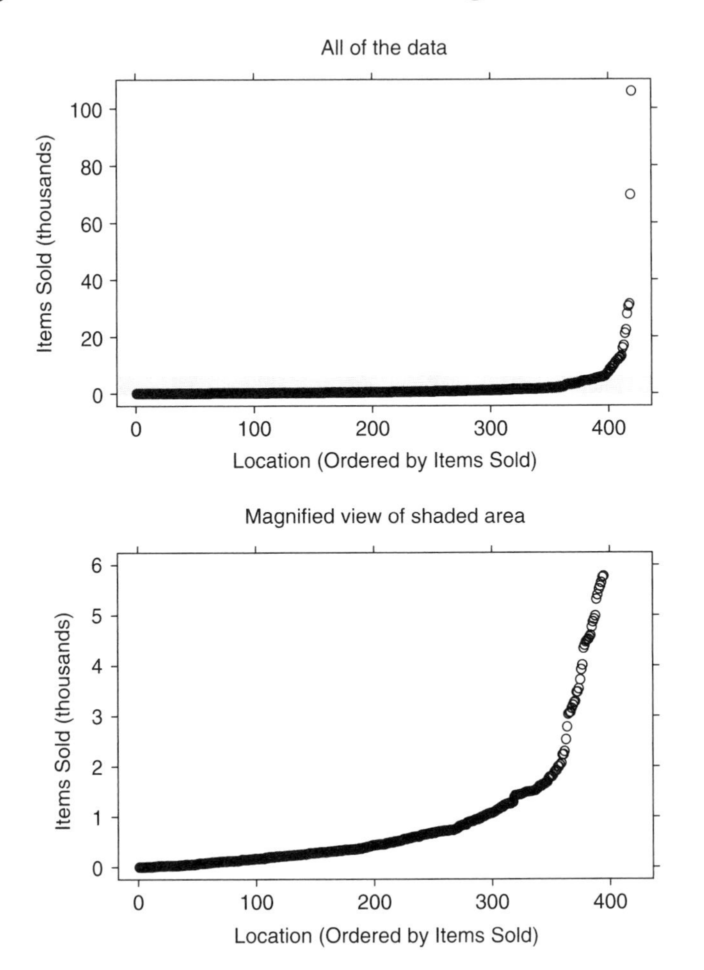

Even a full scale break is not appropriate for audiences who do not read labels carefully and might miss the change of scale. A solution for this audience would be to show all of the data as in the top plot in Figure 7.14, together with a magnified version of most of the data, as in the bottom plot. The high values can be identified in the figure with all of the data (the values above the shaded area), while there is more resolution for the bulk of the data in the bottom figure. Notice that the scales on the horizontal axis are the same, whereas the scales on the vertical axis differ.

Fig. 7.15 Connecting Sides of a Scale Break

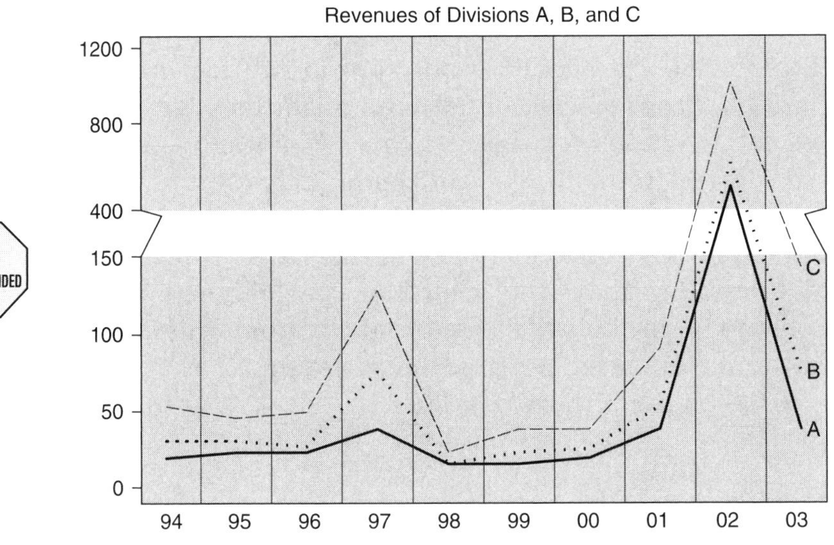

Do not connect numerical values on two sides of a break.

Figure 7.15, similar to those seen in the annual reports of multinational companies, has a forceful scale break: The break is indicated by the area without shading, the break in the grid lines, the jagged scale lines, and the tick marks and labels. Then the break is ignored completely by the lines connecting the two sides of the break. The vertical distance that represents 50 below the break represents 400 above it, so that the line segments from 2001 to 2002 and 2002 to 2003 are meaningless. Connecting the two sides of a scale break creates a very misleading impression.

Fig. 7.16 Temperature Data: Two Scales

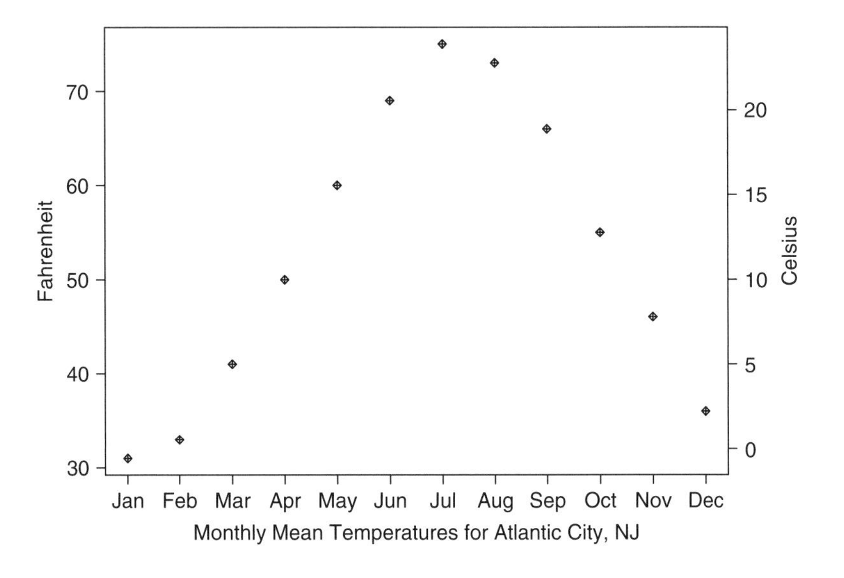

7.5 USING TWO *Y* SCALES

It is often useful to have two scales for one axis.

Figure 7.16 shows the mean monthly temperature for Atlantic City, New Jersey. Some of us are more comfortable with temperature in Fahrenheit; others, with temperature in Celsius. Therefore, both are shown here, Fahrenheit on the left and Celsius on the right. Note that the two scales are equivalent: 32° Fahrenheit corresponds to 0° Celsius. The next example will show two axes that are not equivalent.

Fig. 7.17 Income Data: Gender Gap at All Ages

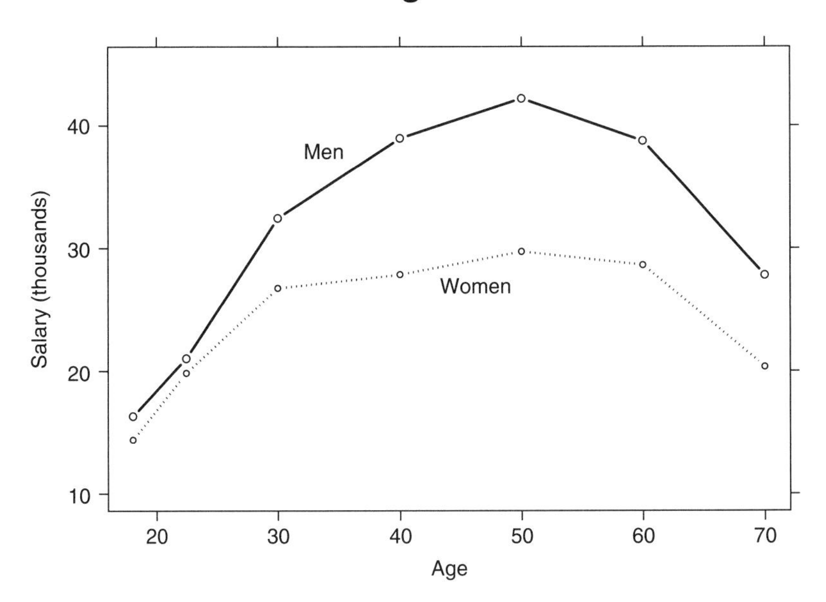

Avoid deceptive double-*y* axes.

Figure 7.17 shows the median salary for men and women taken from the Web page of the Bureau of Labor Statistics. The points plotted are the median usual weekly earnings of full-time wage and salary workers by age and gender, first quarter 2001 averages, not seasonally adjusted, converted to annual figures. It shows clearly that men of all ages earn more than women earn and that the gap is widest around ages 40 to 60. Now suppose that for our purposes we want to hide this fact. The next page does just that, showing how double-*y* axes can be used to mislead.

Fig. 7.18 Income Data: Gender Gap Depends on Age

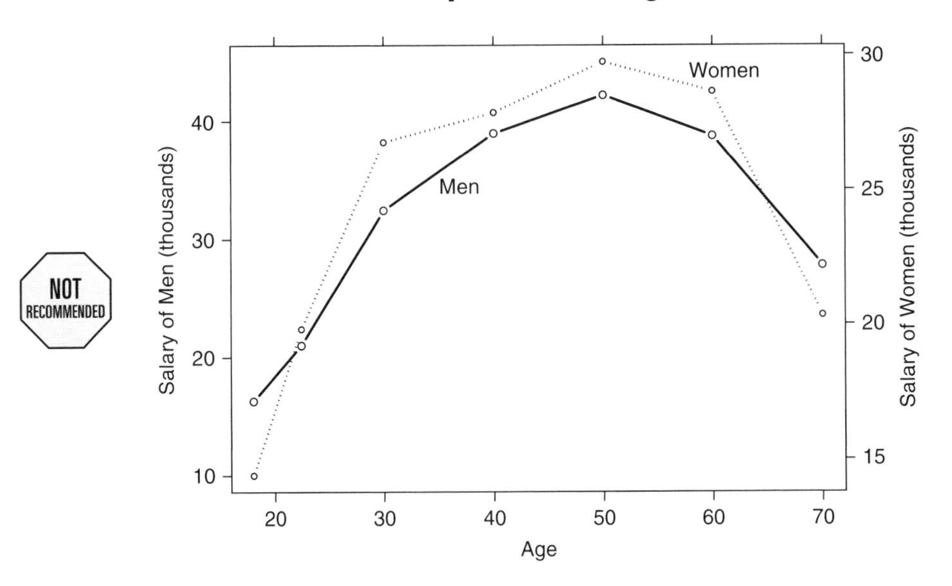

Figure 7.18 shows the salary data with two scales: the salary of men on the scale on the left and the salary of women on the scale on the right. I played with the scale on the right until I got the effect I wanted. The visual impression is that for many ages women earn more than men. Then the graph is finished by using a title that reinforces the deception, as was done in the smoker/nonsmoker example of Wainer (1997). The problem is that the two scales are not the same.

If I had used Excel, I wouldn't have had to play with the axes. The default scales with these data and double-*y* axes in Excel produce a chart that looks very similar to this one.

Fig. 7.19 Blood-Level Data

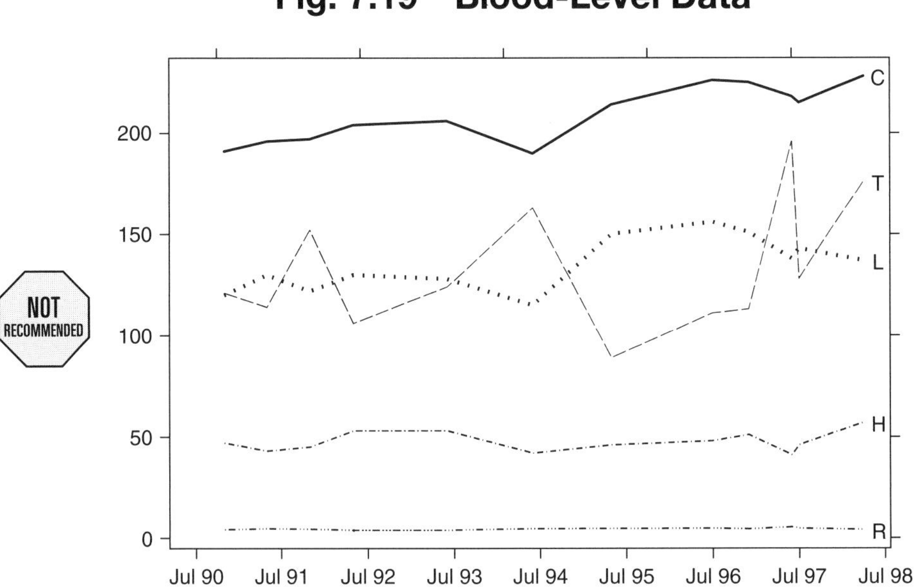

7.6 DATA HIDDEN IN THE SCALES

Choose an aspect ratio that shows variation in the data.

A friend came to me and said that his doctor wanted him to start taking cholesterol medication since his cholesterol had been increasing. He said that he plotted the results of his blood tests and showed me a chart that looked like Figure 7.19. Did I agree that there was a real rise in his cholesterol? There were five line series: one each for cholesterol (C), triglycerides (T), LDL (L), HDL (H), and the ratio (R) of cholesterol to HDL. From Figure 7.19 it appears that there is no variation in the ratio, but that is because his ratio levels are under 5, while his cholesterol levels are around 200. Plotting the two on the same figure hides the variation of the ratio in the scale. Figure 7.19 makes little sense since all of the lines have different units of measurement.

Fig. 7.20 Blood-Level Data: Variation in Ratio

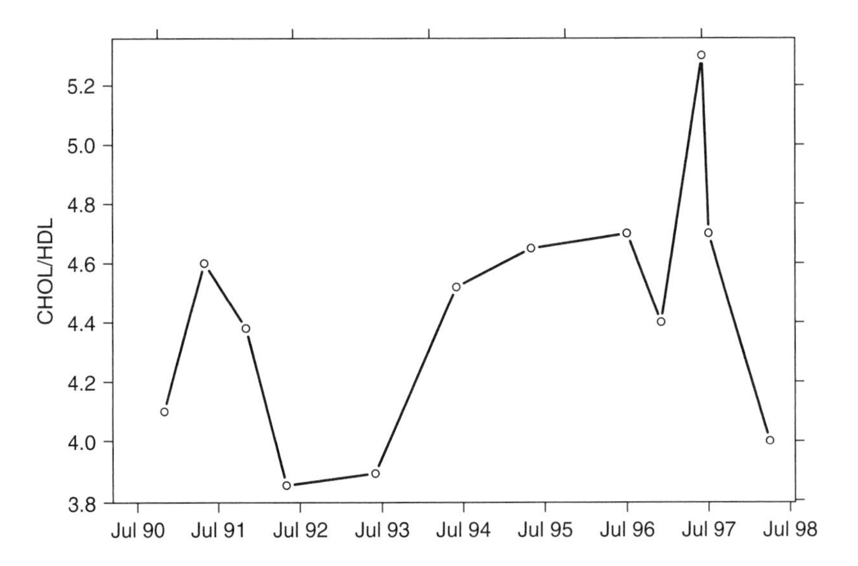

The ratio of cholesterol to HDL is plotted on a scale based on its maximum and minimum values during this time period. We see in Figure 7.20 that the ratio was not at all stable.

Fig. 7.21 Blood-Level Data: Multipanel Plot

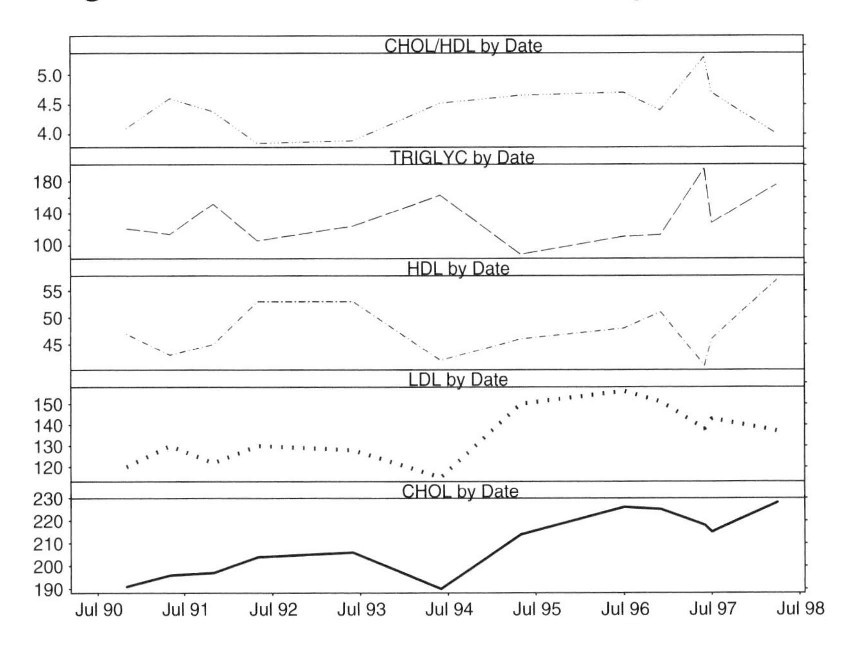

Figure 7.21 plots each of the variables on its own scale. It shows the trend of each of the variables over time. As we just saw, it does not make sense to use the same scale for each panel. We notice that for this particular patient in this time period, HDL and triglycerides appear to be negatively correlated; the triglycerides have peaks at the times the HDL has troughs.

Fig. 7.22 Blood-Level Data: Scatterplot Matrix

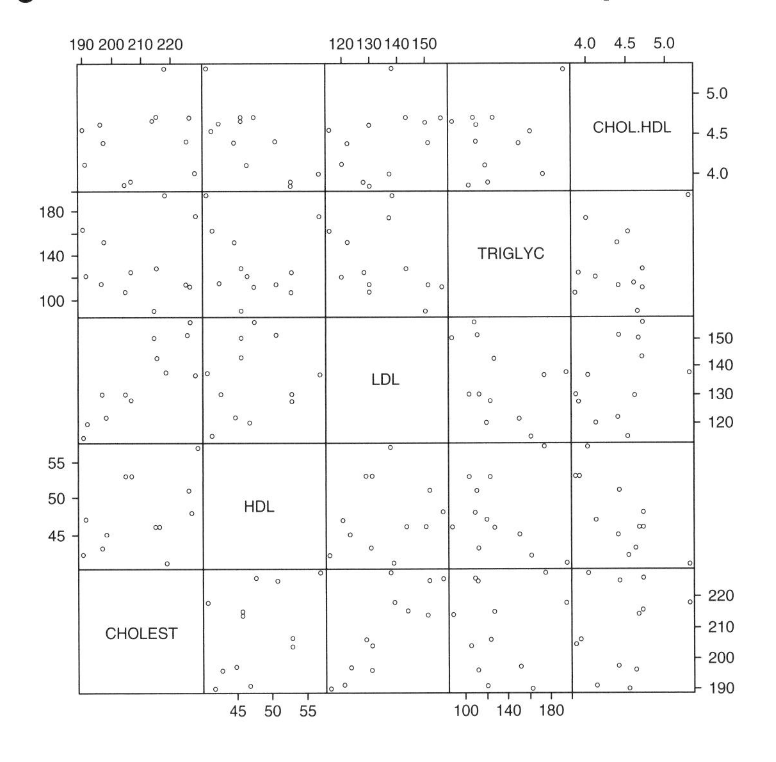

A scatterplot shows the relationship of two variables; a scatterplot matrix shows all pairs of scatterplots. Look at the second row down in the second column from the left of Figure 7.22, showing HDL on the horizontal axis and triglycerides on the vertical axis. We see that HDL and triglycerides do appear to be negatively correlated, but there is one point with high triglycerides and HDL values. The scatterplot matrix enables us to scan for other pairs of variables that appear to be correlated. I have not included the dates to save space, since we already saw the relationship of each variable with time.

Fig. 7.23 Annual Report: No *Y* Scale

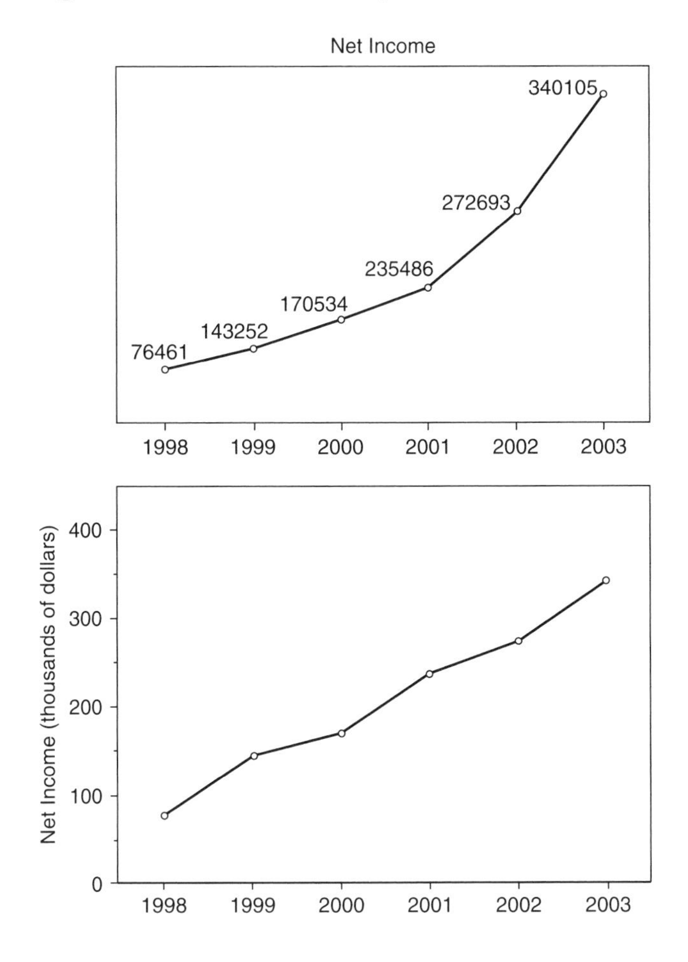

7.7 OTHER PRINCIPLES INVOLVING SCALES

All axes require scales.

In *Designing Infographics*, Eric Meyer (1997) writes: "The more clues to meaning that are supplied elsewhere, the less the need for cluttersome scales." According to this view, a *y*-axis scale line is not needed in the top plot in Figure 7.23 since the data points are labeled with values. The problem with Meyer's position is that not all graphs are drawn accurately. When I see a graph without a scale, I pull out my ruler and measure carefully. I am constantly amazed at how often the figure is not drawn to scale. A scale line with round numbers (such as multiples of 50 or 100) as tick marks helps me to assess whether the numbers look reasonable. Despite the precision implied by the number of digits in the top plot, the numbered labels are not consistent with the plotted points, giving the visual impression of a steeper increase than is actually the case. Scales would help to see this distortion. The bottom plot, with a *y*-axis scale, shows the data (given by the labels of the top plot) drawn to scale.

Fig. 7.24 Prime Lending Rate Data: No *X* Scale

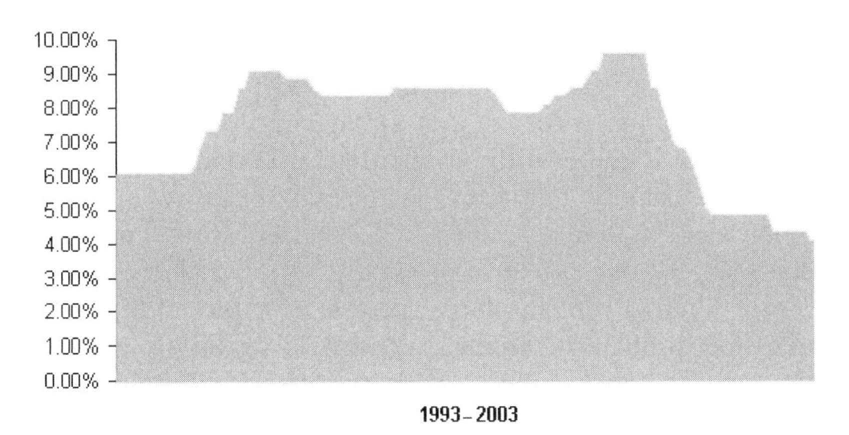

Figure 7.24[1] shows a 10-year trend of the prime interest rate. Although the range of the *x* axis is given, I would find it more useful if there were an *x* axis. Why should the reader have to take out a ruler to determine when interest rates were at a high point? Tick marks on the right as well as the left would make it easier to estimate the rates for more recent years.

The clutter on the *y*-axis labels can be reduced by eliminating the decimal places and saying percent only once: that is, labeling with the digits 1 through 10 and including the axis label "percent."

[1] *http://www.interestonlyloans.com/prime_rate_history.html*, accessed March 4, 2004.

Fig. 7.25 Annual Report Data: Tick Marks at Hard-to-Judge Values

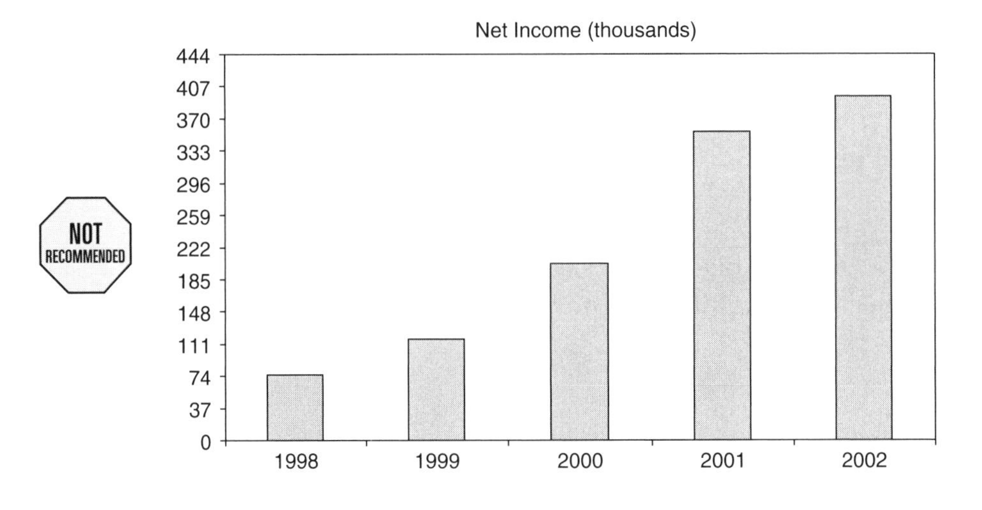

Tick marks should be at sensible values.

It is much harder to judge the values of these bars than if there were a tick mark every $50,000 or $100,000 on the y axis. There are instances in which the most sensible values are not round numbers, but the values in this figure make no sense. A client sent me a graph to review and I suggested changing the tick marks from multiples of 50 to multiples of 60. The reason was that the data represented time measurements in minutes, and most readers recognize 60, 120, 180, . . . as full hours, whereas 100, 200, 300, . . . are more elusive quantities.

Fig. 7.26 IRS Data: Departing from Convention

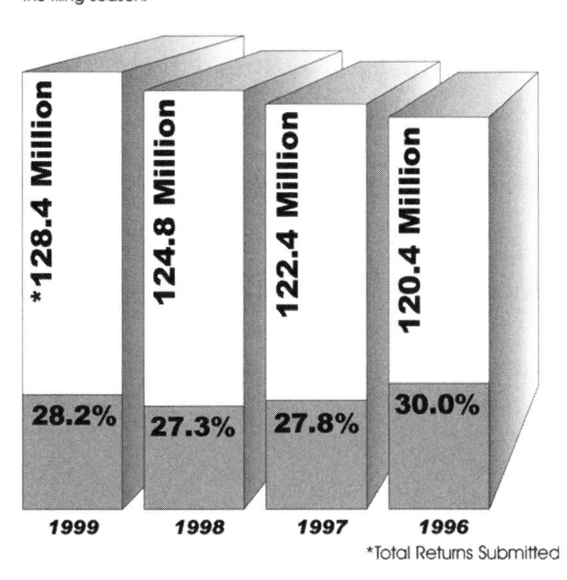

The horizontal axis should increase from left to right and the vertical axis from bottom to top.

In high school we learned that in the Cartesian coordinate system the x axis increases from the origin to the right and the y axis increases upward. Since many readers assume that this is the case and do not read the scale labels carefully, we should stick with that convention. Readers who do notice a reversed scale often think that an attempt was made to mislead them. Figure 7.26 gives the erroneous impression that the number of last-minute income tax filers is decreasing. Time is the most common variable to have its direction reversed, but I have seen this problem with other variables. In one chart the level of education had less than eighth grade on the right, proceeding to graduate degree on the left.

Fig. 7.27 Invoice Data: Filling the Scale-Line Rectangle

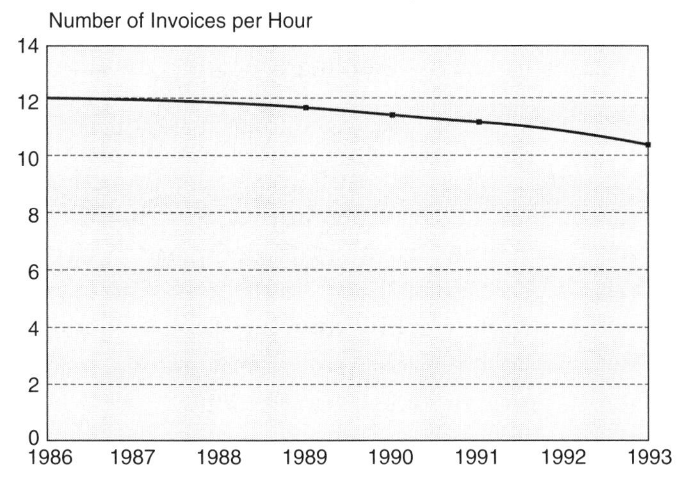

Subject to the constraints that scales have, choose the scales so that the data rectangle fills up as much of the scale-line rectangle as possible.

About 85% of the area of Figure 7.27 contains no information (Texas State Auditor's Office, 1995). All we see are grid lines. Use your judgment to determine whether your audience will read axis labels or whether this waste of space can be eliminated.

Fig. 7.28 Body Mass Data: Unequal Intervals

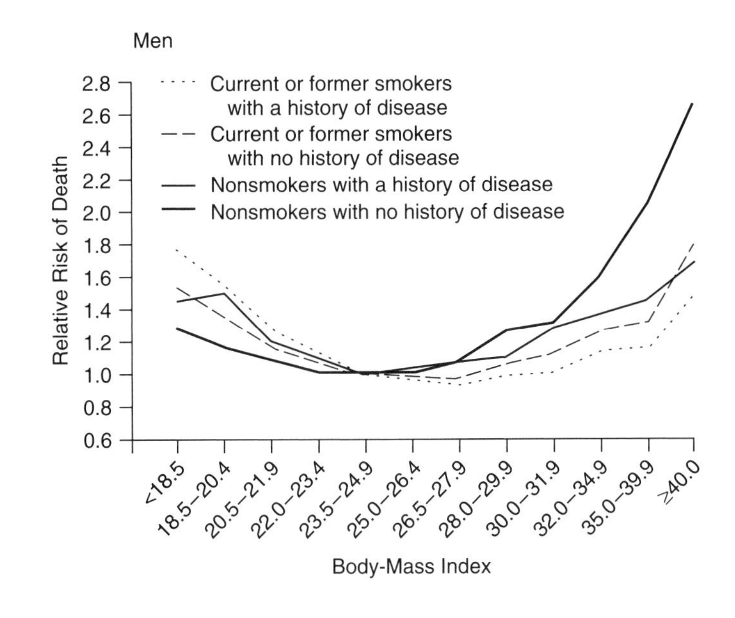

Do not use equally spaced tick marks for uneven intervals on an arithmetic scale.

The *x* axis in Figure 7.28 begins with a body mass index of less than 18.5 (Calle et al., 1999). The next tick mark is 2 units larger, followed by five intervals of 1.5 units, two of 2 units, one of 3 units, one of 5 units, and ends with anything greater than 40. Using these uneven intervals as if they were the same distorts the shape of the curves. It is another form of not drawing the data to scale.

This problem occurs frequently with Excel users who use line charts for unequal time intervals. In Section 9.3.2 I continue this discussion for Excel users and provide another example.

Fig. 7.29 Body Mass Data: Comparing Panels

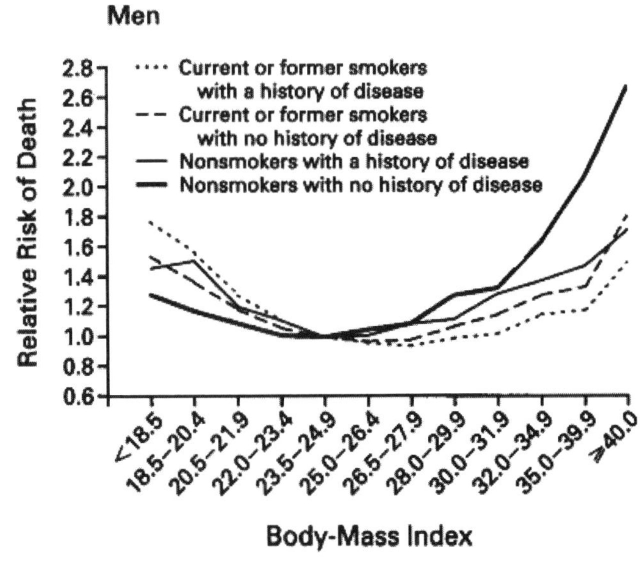

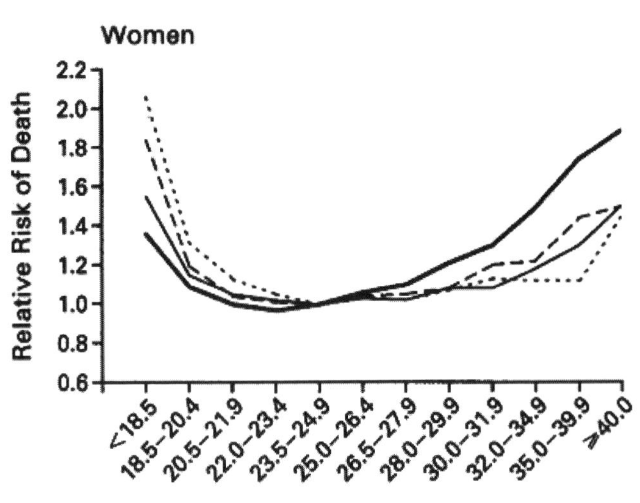

Choose appropriate scales when data on different panels are compared.

Figure 7.28 is part of the two-panel display shown in Figure 7.29 (Calle et al., 1999). Not only are the horizontal axes not drawn to scale, but the comparison of males to females with the vertical axes is misleading since it uses different scales for men and for women. When comparisons are to be made, it is preferable to use the same scale for both figures. That is easy to do in this example.

Sometimes this is not possible because the levels of the various sets of data vary by too much; using the same scale would hide the variation of a variable. A second choice is to use the same number of units per inch. Even this is not possible for some sets of data, since a scale on one of the panels could become too small. In that case it is useful to add rectangles that show the relative scales, as was done in Figure 4.6.

Fig. 7.30 Body Mass Index: Corrected

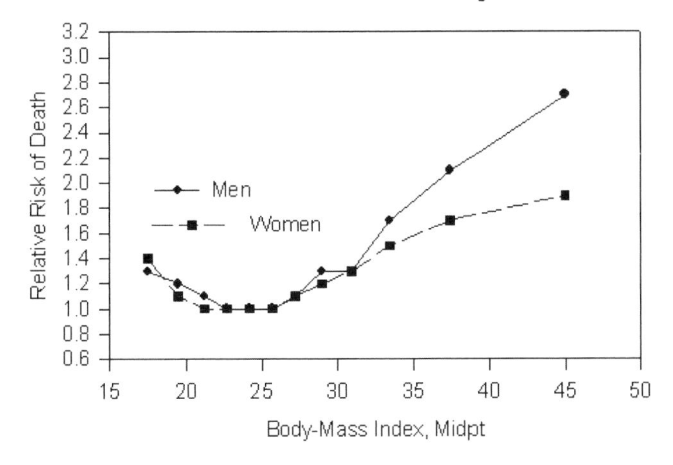

Flaster (1999) redrew these figures correctly. Note in Figure 7.30 that the *x* axis now has evenly spaced tick marks for the same interval of body mass index, and the *y* axes of both men and women are the same. The figures now look more like the letter J, whereas before they looked more like a U.

If our primary interest is a comparison of the relative risks for men and women, showing both men and women in the same plot makes comparison easier. However, if our primary interest is comparing the relative risks for the four groups of the same gender, corrected versions of Figure 7.29 are preferred. Once again, showing the data in more than one way increases our understanding.

SUMMARY

Choose scales wisely, as they have a profound influence on the interpretation of graphs. Not all scales require that zero be included, but bar graphs and other graphs where area is judged do require it. Be careful using double-*y* axes so you do not mislead your audience.

8 Applying What We've Learned: *Before* and *After* Examples

In this chapter I apply the principles and graph forms we just learned to real-life examples of charts and graphs. The first example is a grouped bar chart of the yearly prices of gold as well as an alternative presentation of the data using high-low charts. The second is a group of separate charts that are more effective when unified. The third shows data presented by a radar chart redrawn with a custom multiway chart. The next example shows trends much more clearly in a trellis display than with multiple pie charts or a stacked bar chart. Finally, tabular data are redrawn using multiway dot charts to emphasize patterns in the data.

Creating More Effective Graphs, by Naomi B. Robbins
ISBN 0-471-27402-X Copyright © 2005 John Wiley & Sons, Inc.

The *before* examples are not deceptive and do not exhibit some of the serious errors seen in Chapter 6. They just don't communicate their information as effectively as certain alternatives do. In some cases, the *after* examples are drawn using graphs available in common software as well as with my choice of software. But not all of the examples have simple solutions that can be drawn easily with Excel or other common packages. I hope that the reader realizes that there are options with Excel and other everyday software, that one set of choices might produce much clearer graphs than another, but that there still are limitations to what can be accomplished when one is restricted to these programs.

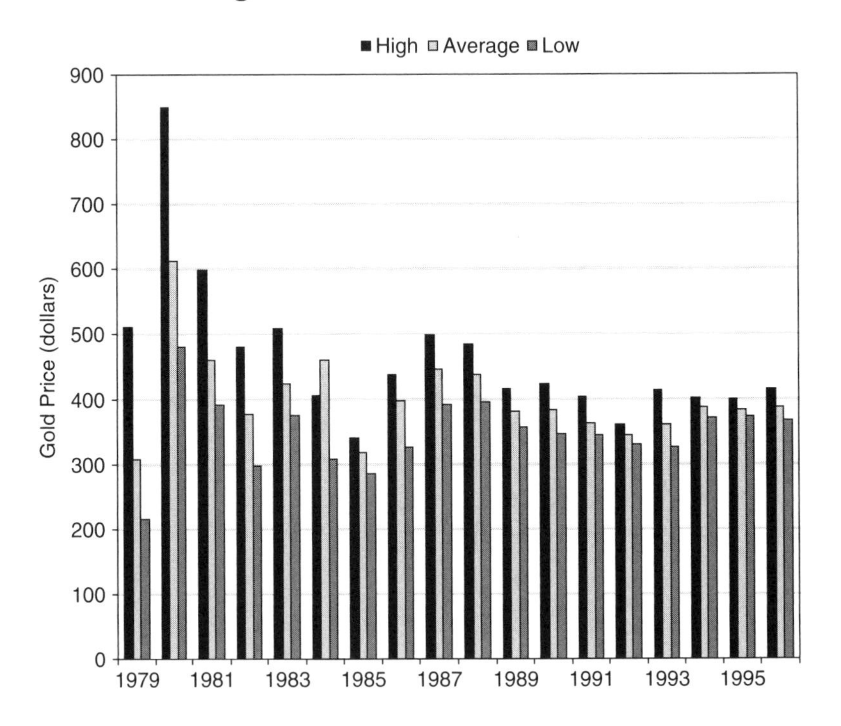
Fig. 8.1 Gold Price Data
High
Average
Low
Gold Price (dollars)
900
800
700
600
500
400
300
200
100
0
1979
1981
1983
1985
1987
1989
1991
1993
1995

8.1 GROUPED BAR CHART

The *grouped bar chart* in Figure 8.1 shows the yearly high, average, and low prices of gold for the years indicated. I'm often shown grouped bar charts and asked for alternative ways to show the data since the designer sensed that it did a poor job of communicating information. In Chapter 2 we learned that trends in grouped bar charts are difficult to perceive since too much extraneous information is nestled between the relevant readings. For example, it is difficult to follow the average gold price since the high and low bars on either side of the average interfere with our ability to follow fluctuations in the average from 1979 to 1996.

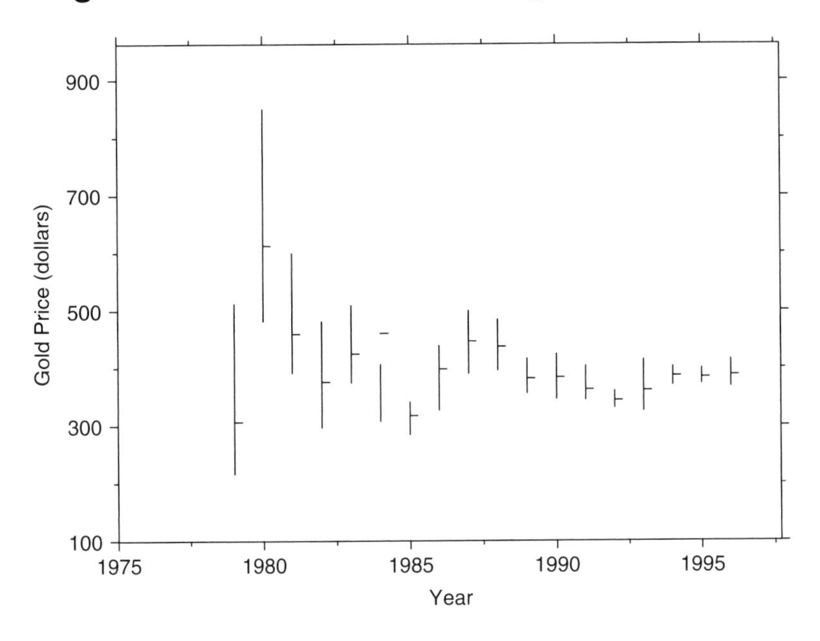
Fig. 8.2 Gold Price Data: High-Low Chart
Gold Price (dollars)
900
700
500
300
100
1975
1980
1985
1990
1995
Year

The gold price data are plotted again in Figure 8.2 using a high–low chart. Each line segment goes from the low to the high value for the given year, with the average indicated by a short horizontal line. This figure is less cluttered, making it easier to see trends, to get a sense of the range of the data for each year, and to compare year-to-year variations. We notice immediately that the average for 1984 is higher than the high, probably indicating a mistake in the data.[1] Thus, a more effective graph of the data pointed out an error in the data as well as displaying the trends more clearly.

[1] Data from *http://goldinfo.net/londongold.html* confirm that the average presented for 1984 was a typo.

Fig. 8.3 Task Data

Average Time Spent on Stereo Tasks: Original vs. Revision

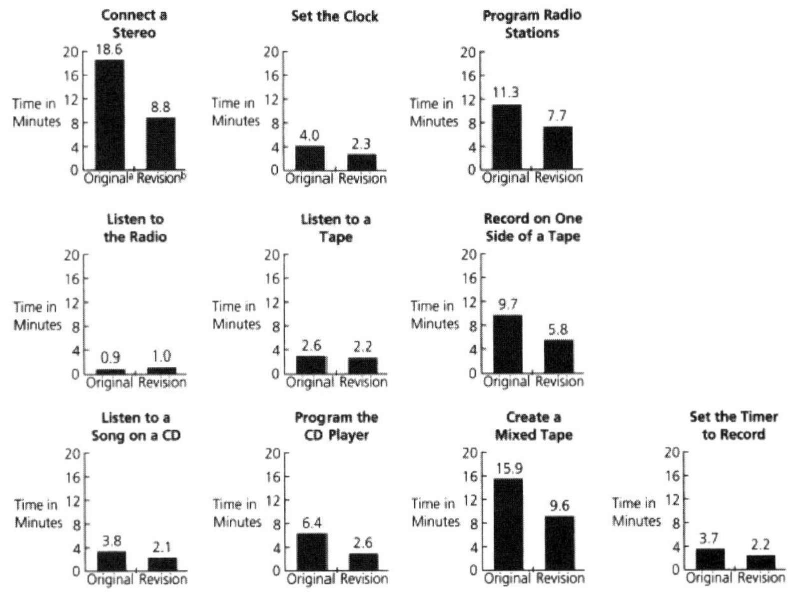

Note. Values represent participants' mean time per task.

[a] 17 participants (three English-only speakers, five French bilinguals, four German bilinguals, five Spanish bilinguals) completed 10 tasks using the original instruction guide.

[b] 12 participants (three English-only speakers, three French bilinguals, three German bilinguals, three Spanish bilinguals) completed 10 tasks using the revised instruction guide.

8.2 TEN SMALL GRAPHS

In her important book *Dynamics in Document Design: Creating Texts for Readers*, Schriver (1997) states that one of the goals of her book is "to demonstrate advantages of taking the reader's needs seriously." In her section on legibility and quantitative graphics, she claims that research from psychology suggests "that it is important to keep the reader's cognitive load as light as possible."

Schriver describes an experiment in which users were given technical manuals with instructions for installing VCRs and stereos. The users were watched, timed, and interviewed while performing these tasks. Technical communications experts rewrote these manuals and a new round of experiments took place. Figure 8.3 shows the times to complete the stereo tasks using the original and final revised manuals.

Fig. 8.4 Task Data Unified

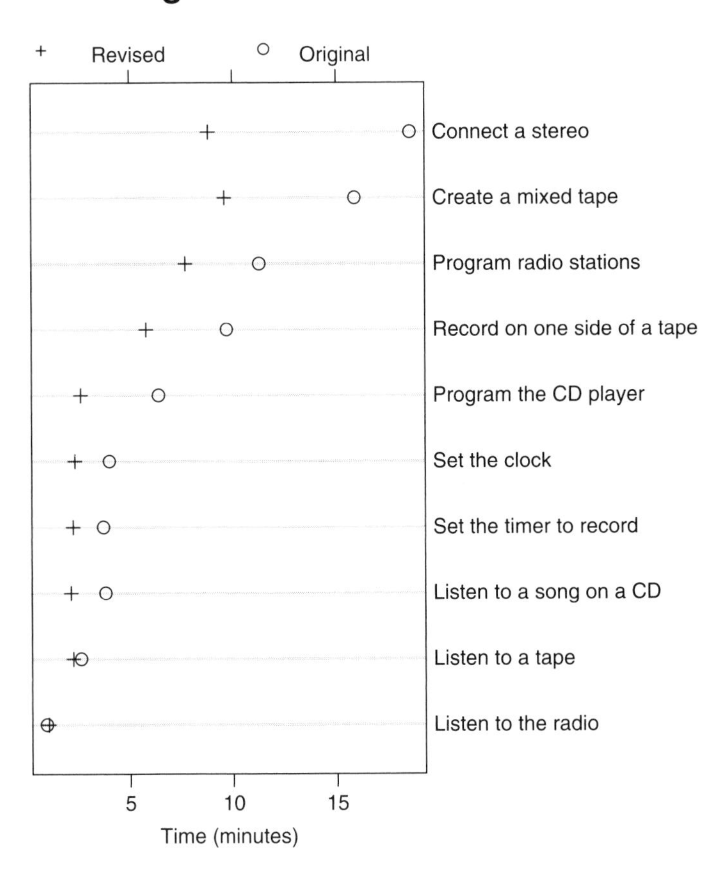

Which task took longest to complete? Which task showed the most improvement? To answer questions such as these, the reader must scan the page and perform mental calculations. Figure 8.4 unifies the data. The tasks are ordered by the average time it took to perform them using the original manuals. It is certainly easier to compare tasks with this figure.

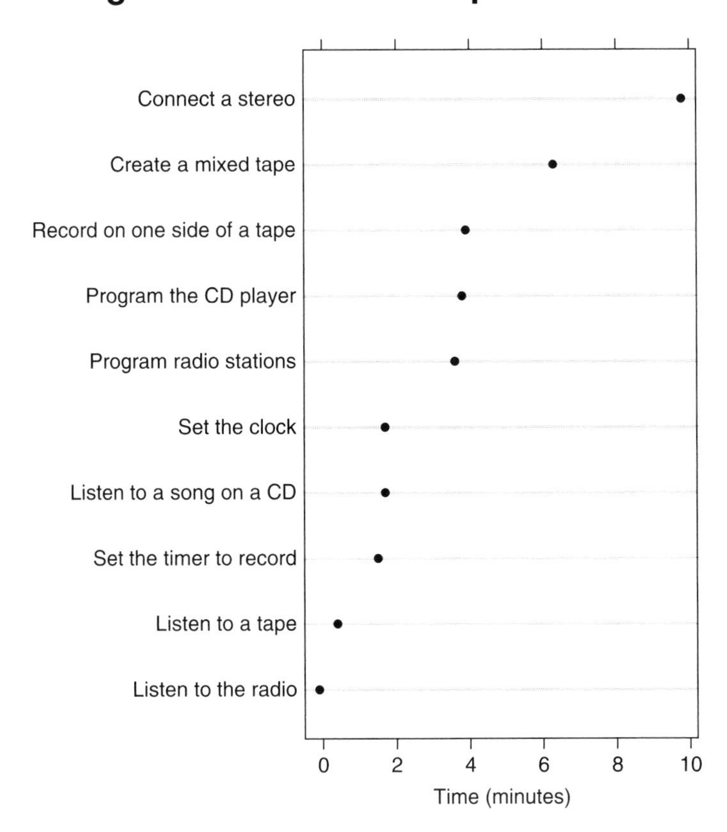
Fig. 8.5 Task Data: Improvement
Connect a stereo
Create a mixed tape
Record on one side of a tape
Program the CD player
Program radio stations
Set the clock
Listen to a song on a CD
Set the timer to record
Listen to a tape
Listen to the radio
0
2
4
6
8
10
Time (minutes)

Figure 8.4 still emphasizes the *before* and *after* times. If we are interested in the improvement, we should let the computer do the calculations and plot the improvement rather than the *before* and *after* times, to keep the load on the reader lighter. Also, for any task in Figure 8.4, the reader has to realize that the lower time is preferable. That step is no longer necessary in Figure 8.5.

Fig. 8.6 Task Data: Percent Improvement

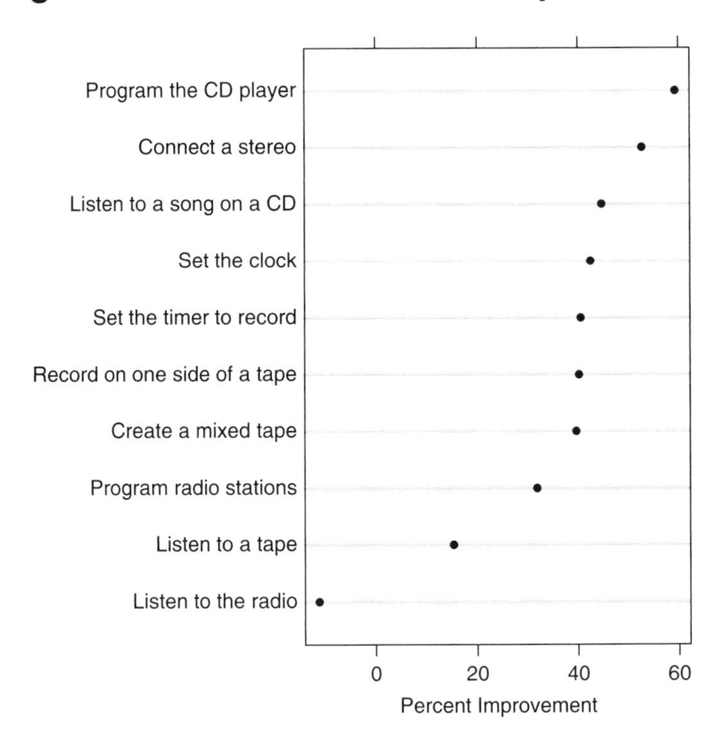

Percent improvement makes more sense than absolute improvement, since some tasks, such as listening to the radio, have little room for improvement since the times are so short to begin with. In Figure 8.6 we plot the percent improvement.

Which of these figures is best to use? This is not a graphical question. It depends on the audience and the information that the audience needs. It also depends on the information that the designer chooses to communicate. The variable plotted should be the variable that the reader needs to know without the necessity of mental calculations.

Fig. 8.7 Task Data: Bar Chart

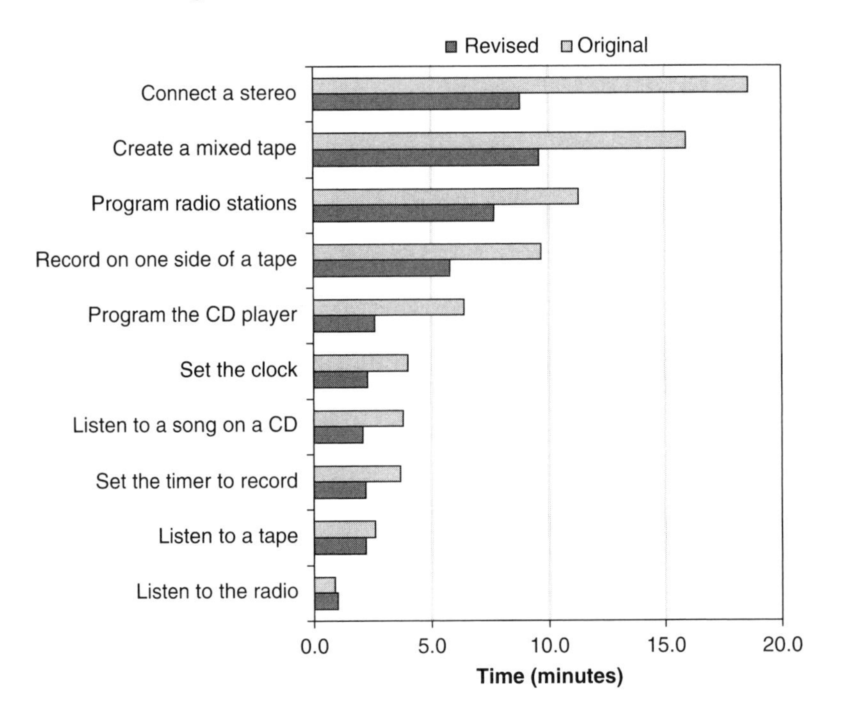

Revisions to the task data use dot plots. I like dot plots. Section 9.3.3 describes a macro for drawing dot plots with Excel. However, the principles applied can be achieved with any graphing software, since they all include bar charts. Figure 8.7 shows how we can use bar charts to unify the information by using one graph instead of ten, and order the information by the length of time needed to complete the tasks. Charts showing improvement and percent improvement can be drawn using bar charts as well. Combining 10 small charts into one eliminates the need to repeat *time in minutes, original*, and *revision* 10 times each, thereby reducing clutter.

8.3 RADAR CHART

The Association of Research Libraries (ARL) has sponsored initiatives to replace input measures of library quality, such as size of collections, with outcome measures, such as quality of service. One of these initiatives has been the LibQUAL + (tm) project, a survey of users of the library. The 2002 LibQUAL + (tm) survey consists of 25 questions in four categories: affect of service, library as place, personal control, and access to information. There are four user groups participating in the survey: faculty, graduate students, undergraduate students, and other. The participants were asked to specify the importance to them of each question by specifying the desired level of service on a scale of 1 to 9. They were also asked to specify the minimum acceptable level of service and the perceived level of service for each question.

The researchers used a radar chart to summarize the results of the survey; the original radar chart appears in color. Figure 8.8 shows the results for faculty surveyed from four-year institutions of higher learning. Each radial axis represents one of the 25 questions with 0 at the center of the figure and 9 at the circumference. The shaded area shows the range from minimum acceptable to desired when the perceived falls in this range, and the range from perceived to desired when the perceived value is less than the minimum acceptable (e.g., question 3). Question 5 is more important to this user group than question 23 since the shaded range is closer to the circumference. For more information on the ARL Statistics and Measurement Program, see *http://www.arl.org/stats/*. The wording of the 25 questions is available from the ARL notebooks (ARL, 2002, pp. 106–107).

Fig. 8.8 ARL Library Survey: Radar Chart

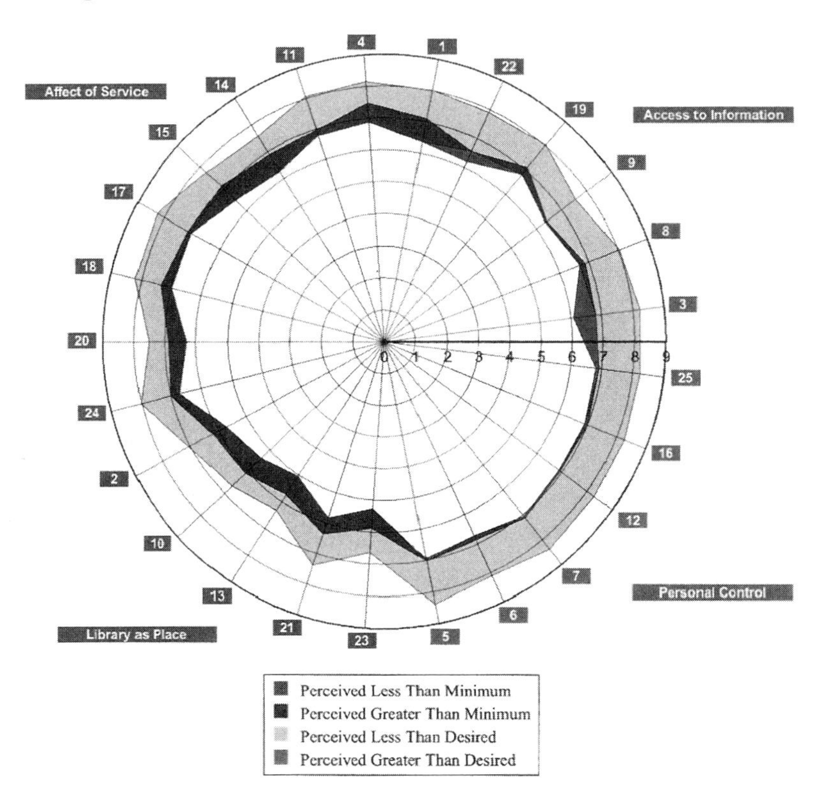

Source: ARL Notebook 2002 LibQUAL + (tm) Results (Washington, DC: Association of Research Libraries, 2002). 4.5.1: Item Summary for 4-Year Institution—Faculty [radar graph], p. 46. Retrieved January 20, 2003 from *<http://www.libqual.org/documents/admin/ARLNotebook111.pdf>*.

A careful reading of the radar chart shows a legend item, *Perceived Greater Than Desired*, that does not appear in the figure. These charts are prepared in an automated fashion for a large number of libraries, and not every legend category occurs in every chart. Many readers find these radar charts difficult to interpret since they are read so differently from graphs that they are used to. Figure 8.9 shows an alternative presentation of the data that is easier to understand.

Fig. 8.9 ARL Library Survey: Variation of a Trellis Plot

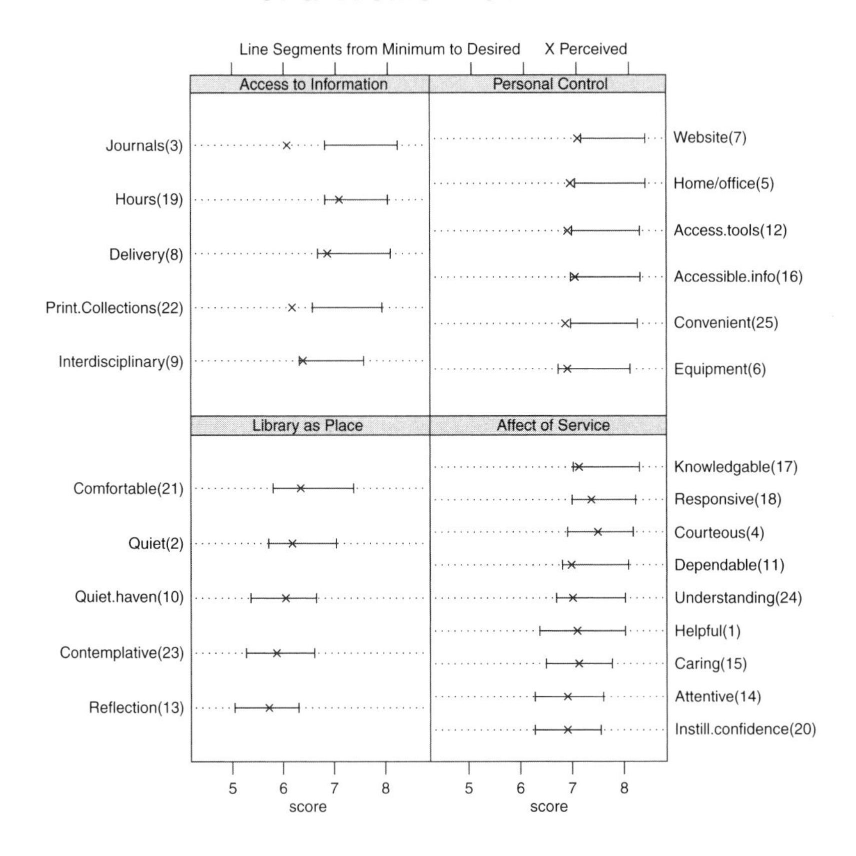

An alternative to the radar chart is shown in Figure 8.9 using the principles of trellis graphics and multiway charts. The line segments show the range from the mean minimum acceptable level to the mean level desired for each question. Note that there is room to refer to the question by keyword instead of the question number used in the radar chart. The mean perceived level is shown by an X. If two questions have the same desired level, the one with the higher minimum acceptable level is more important to the user, so we define importance by the average of desired and minimum. Within each group the questions are ordered by importance. The categories are also ordered by importance, with ''library as place'' the least important to this faculty user group, and ''personal control'' the most important.

Fig. 8.10 ARL Library Survey: Comparison of User Groups

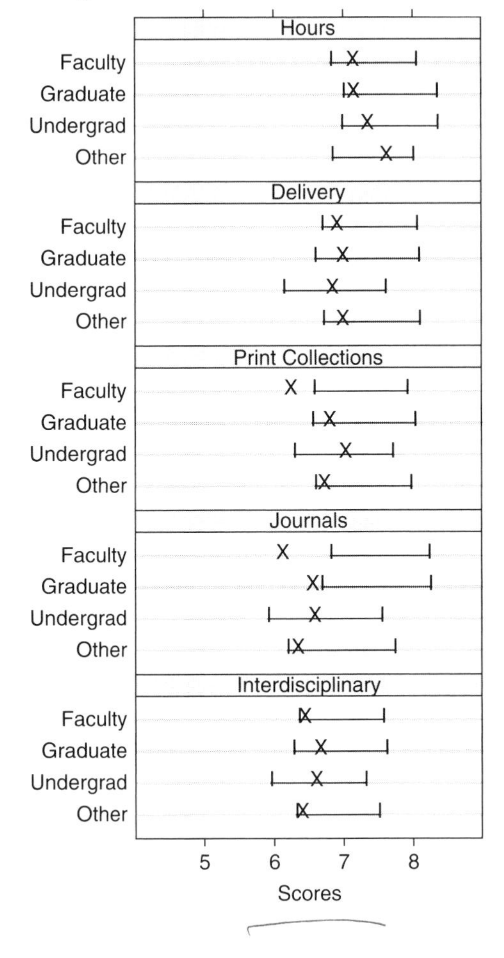

Figures 8.8 and 8.9 show the results for the faculty, emphasizing the different responses to the 25 questions by this user group. Similar charts are available for the other user groups. In Figure 8.10 we emphasize the different responses of the four user groups to a given question, showing the questions for the access to information category. Similar charts are available for the other categories. Notice that although faculty considered that a complete run of journal titles was the most important question in this category, convenient hours were more important to all users combined. The faculty members surveyed identified a problem with print collections and journal titles, whereas the undergraduates were satisfied with these items. The questions are ordered by importance, with the most important on the top.

Fig. 8.11 Car Production: Multiple Pie Charts

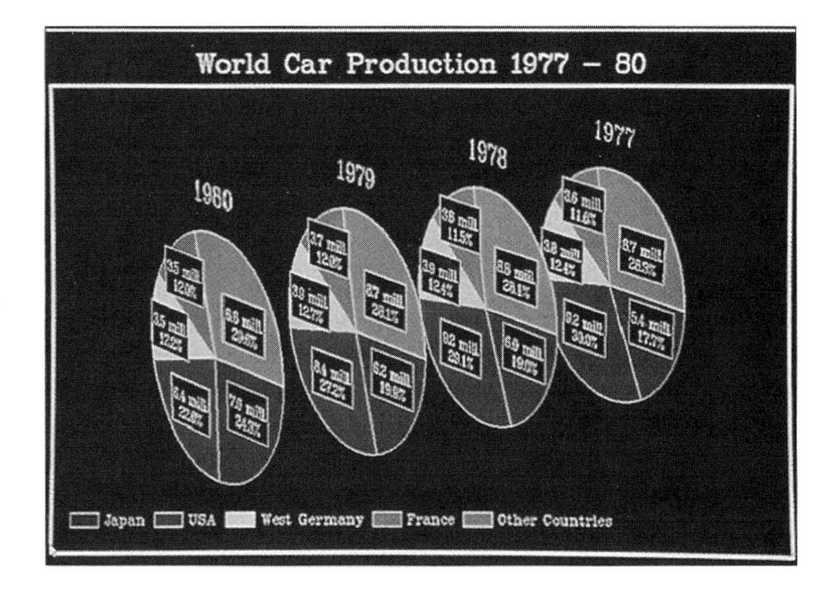

8.4 MULTIPLE PIE CHARTS

Brown et al. (1995) include a color version of Figure 8.11 in their "high-quality presentation graphs," although they do admit that "angles can sometimes be hard to compare." The figure is entitled "A Projected Pie Chart for World Car Production." It compares the number and percentage of cars produced by each of the four major producing countries and a fifth category, other, which combines other significant automobile-producing nations. There are a number of problems with Figure 8.11. First, it is very difficult to read. Making comparisons from multiple pie charts is always a difficult task. The perspective is intended to ease this task. Does it work for you? Years are usually shown from left to right, so this presentation may mislead the reader.

Fig. 8.12 Car Production: Stacked Bar Chart

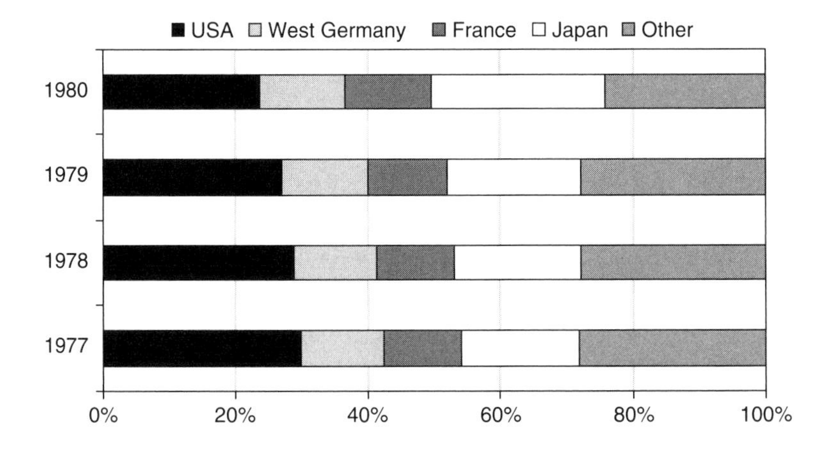

Spielman (2003) also criticizes multiple pie charts and questions whether the reader can "track who is gaining or losing market share and at what rate." He redraws the data with a stacked bar chart. This black-and-white stacked bar chart suggests the look of his colored chart. I agree that it is a huge improvement over Figure 8.11, but we learned in Chapters 2 and 3 that it is difficult to compare lengths that do not have a common baseline. If limited to everyday software and if color were an option, I would prefer multiple line graphs.

Fig. 8.13 Car Production: Trellis Panels

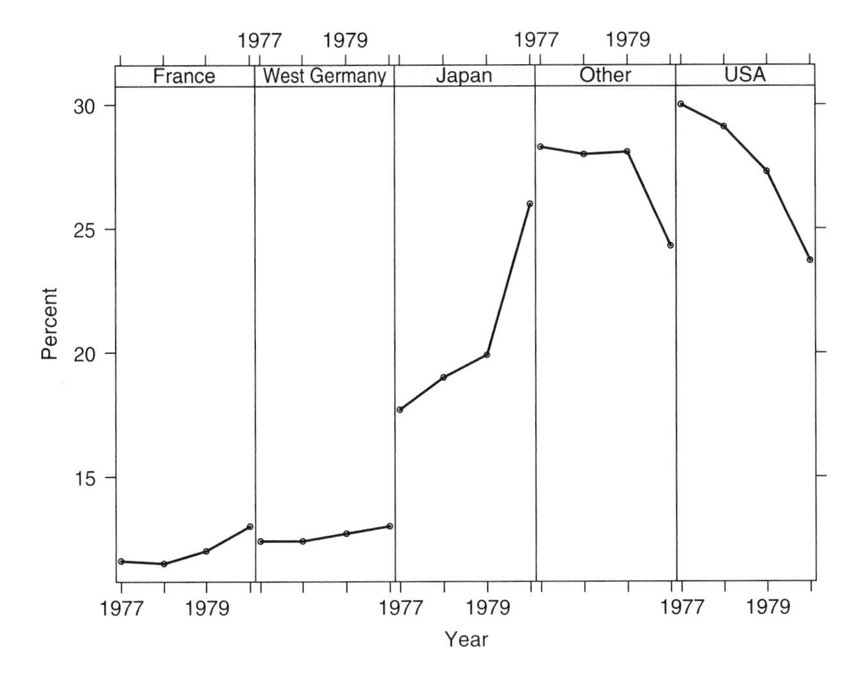

Figure 8.13 is my preference for showing trends in market share. It shows clearly who is gaining and who is losing market share. Although it does not show the totals, I checked the data carefully and made sure that the totals for each year are 100%. The stacked bar chart shows the trends well for those countries with large changes, such as Japan and the United States, but is less clear for France and West Germany, where the trends are less pronounced. Another advantage of Figure 8.13 is that it is clear without the need for color.

8.5 TABLES

Tables provide an effective way of displaying exact numbers. Graphs provide patterns and trends; tables provide the details. Put another way, graphs are for the forests and tables for the trees. But large tables cannot be seen on screens and have no place in overheads, slides, or laptop presentations. The table below, showing 2000 Census data of the population of various ethnic groups in some counties in the New York metropolitan area, comes from a PowerPoint presentation (private client, 2003).

County	Bronx	Kings	New York	Queens	Richmond	Nassau
White	194,000	855,000	703,000	733,000	317,000	986,000
Latino	645,000	488,000	418,000	556,000	54,000	133,000
Black	415,000	845,000	233,000	420,000	40,000	129,000
Asian American	38,000	184,000	143,000	392,000	24,000	62,000
All others	40,000	93,000	39,000	128,000	9,000	24,000
Total	1,333,000	2,465,000	1,537,000	2,229,000	444,000	1,335,000

The first problem is that the audience cannot read it. The second is that it doesn't highlight the points the presenter wants to make. Even if the audience could read it, they wouldn't know what to focus on.

Suffolk	**Westchester**	**Rockland**	**Bergen**	**Hudson**	**Passaic**	**Total**
1,118,000	592,000	205,000	638,000	215,000	252,000	6,807,000
149,000	145,000	29,000	91,000	242,000	147,000	3,098,000
92,000	123,000	30,000	43,000	73,000	60,000	2,504,000
34,000	41,000	16,000	94,000	57,000	18,000	1,104,000
26,000	23,000	6,000	18,000	22,000	12,000	442,000
1,419,000	923,000	287,000	884,000	609,000	489,000	13,955,000

Fig. 8.14 Population Data by Race

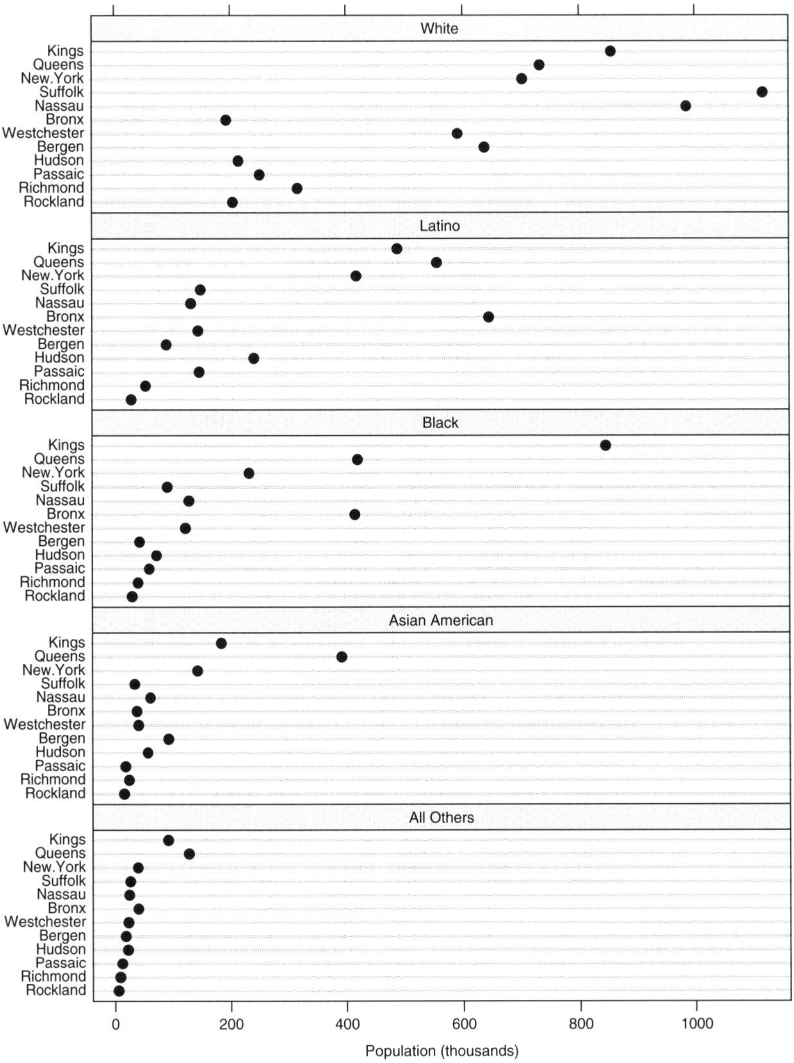

White

Latino

Black

Asian American

All Others

Kings
Queens
New.York
Suffolk
Nassau
Bronx
Westchester
Bergen
Hudson
Passaic
Richmond
Rockland

0 200 400 600 800 1000

Population (thousands)

Once again we use trellis displays and multiway charts to present the data. For each race, the population for each county appears on the horizontal axis with the counties ordered by total population. The races also appear ordered by population. Suppose that you are interested in which county is most diversified or where the greatest concentration of Asian Americans lives. Look at the table and then at Figure 8.14. Where is it easier to find the answers to these questions? The table is useful for finding the exact population of counties; Figure 8.14 points out the patterns in the population statistics.

Fig. 8.15 Population Data by County

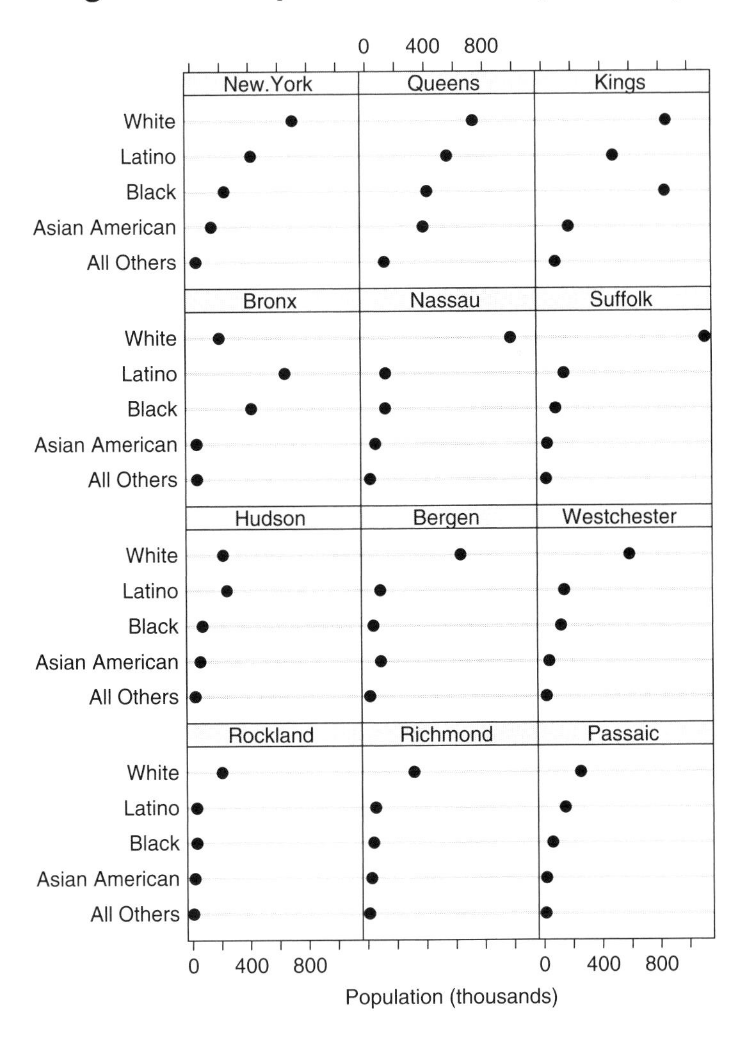

Different presentations of data emphasize different aspects of the data. Multiway charts enable us to see the data in a variety of ways. In Figure 8.15 we emphasize the counties, whereas in Figure 8.14 we emphasized the races. It is helpful to see the data both ways.

SUMMARY

The methods and principles of this book were used to redraw several real figures. The number of times that dot charts and multipanel trellis displays were used in the *after* figures demonstrates their usefulness.

9 Some Comments on Software

Several types of software produce graphs, including statistical software, spreadsheets, drawing programs, and presentation software. Reviews of specific packages, as well as the platforms they run on, appear on the Web.[1]

9.1 STATISTICAL SOFTWARE: S LANGUAGE

The S language was used to draw most of the graphs in this book. Many consider the graphics of S to be the best of statistical software. There are two major implementations of the S language: S-Plus and R. S was developed at Bell Laboratories. An enhanced and supported commercial version of the S language, S-Plus, is available from the Insightful Corporation. Graphs may be produced through either the command language or the graphical user interface. Although the graphical user interface is easier to learn and use, I prefer

[1] All software mentioned in this chapter runs on Windows. Most run on other platforms as well.

Creating More Effective Graphs, by Naomi B. Robbins
ISBN 0-471-27402-X Copyright © 2005 John Wiley & Sons, Inc.

the command language since with it you can save code (sequences of commands) and thus reproduce what you've done later. I find that I forget how I produced certain effects and am referring back constantly to code used previously.

R is an open-source system written by a team of volunteers (R Development Core Team, 2003) and (Ihaka and Gentleman, 1996) which is freely available (*http://www.r-project.org*). R is run using a command language. Since it is free software, any of the graph forms recommended in this book can be created with no expense for software. Since S-Plus and R are programming languages, you have the option of creating graph types even if they don't come with the package. For example, I drew the mosaic plot shown in Figure 5.10 using S-Plus with code from Jay Emerson's Web site (Emerson, 1998).

A companion book to this one is Murrell (2005), which contains documentation for graphics using R. One way to learn to draw graphs with R is to find a figure similar to what you want in Murrell, check his code for that figure, and then look up the commands he used to tweak them for your situation.

9.2 DRAWING PROGRAMS: ILLUSTRATOR

Graphic artists and information designers tend to use illustration software such as Illustrator. Although Illustrator and PowerPoint have tools for creating a graph automatically from a set of data, other popular design software, such as Photoshop and Quark, do not. To create a graph, each line, box, and shape needs to be "drawn" on the screen as if with pencil and paper. To enter data in Illustrator or PowerPoint you can cut and paste from other software, such as Excel. Although PowerPoint is more user-friendly, Illustrator provides many more of the design options used by seasoned graphic designers. Some graph designers use other programs to generate graphs and then fine-tune them with Illustrator.

9.3 SPREADSHEETS: EXCEL

Surely the most common software for producing graphs is Excel. Many users rely on the default settings, which, unfortunately, often produce cluttered, confused, or even misleading graphs. However, by paying attention to the specifics of the data at hand, the user can readily learn which settings are important to change in order to create a much more effective graph.

Excel Charts by Walkenbach (2003) is absolutely required reading for Excel chart users. The book explains how to produce graph forms such as box plots, which at first seem impossible to draw using Excel. On the following pages I provide some tips for Excel users, to minimize clutter and confusion.

Fig. 9.1 Moving an Axis in Excel

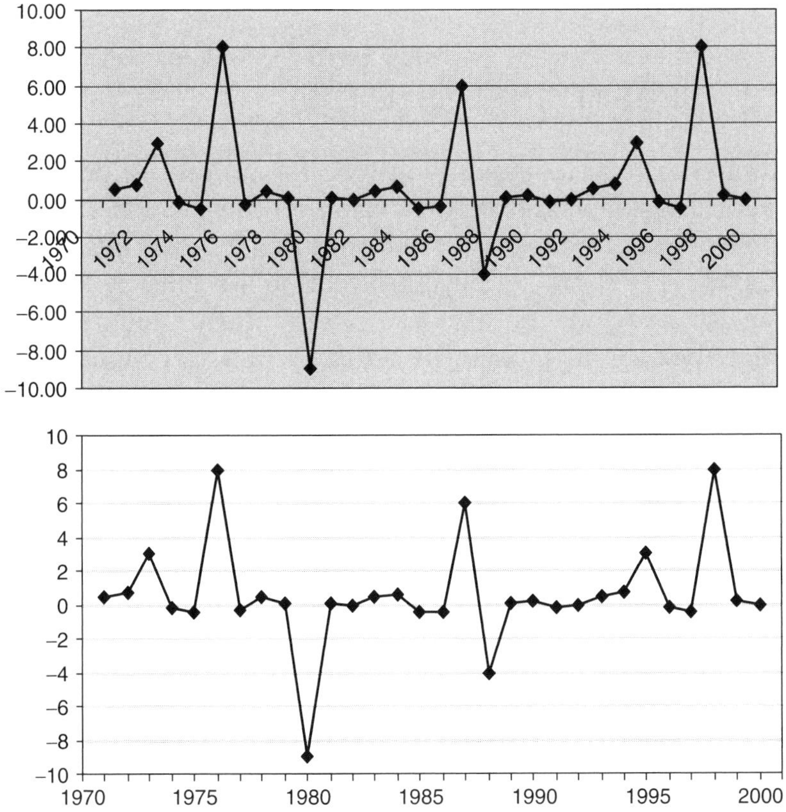

9.3.1 Moving an Axis in Excel

The top of Figure 9.1 shows a line chart produced by Excel using all of the default settings for this chart type. Notice that the horizontal axis falls right in the middle of the data points. As a result, the tick marks and tick mark labels interfere with our view of the data. To move the axis down, right-click on the vertical axis, choose *format axis*, choose the scale tab, and change the value for "Category (X) axis Crosses at." The bottom figure shows the effect of this change along with some formatting changes that make the data stand out better.

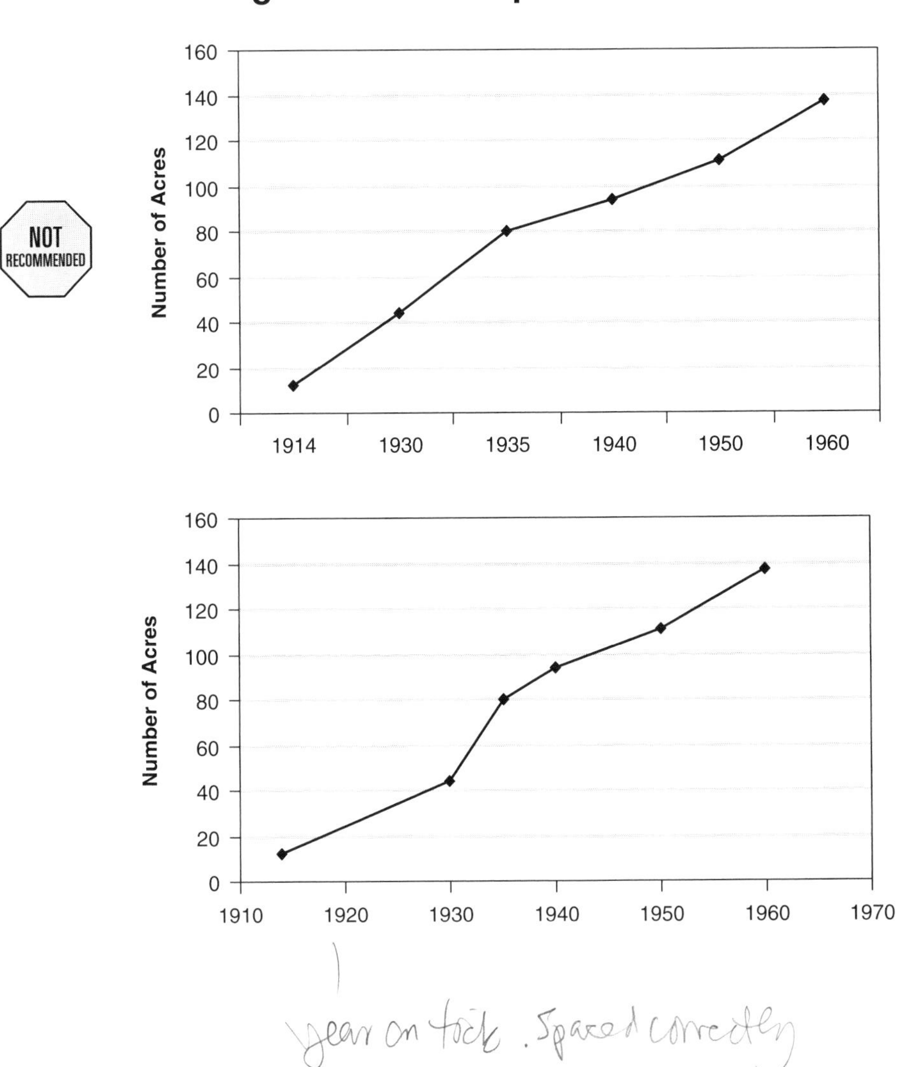
Fig. 9.2 Scale Tip for Excel Users
NOT
RECOMMENDED
Number of Acres
160
140
120
100
80
60
40
20
0
1914
1930
1935
1940
1950
1960
Number of Acres
160
140
120
100
80
60
40
20
0
1910
1920
1930
1940
1950
1960
1970
year on tick. Spaced correctly

9.3.2 Line Charts with Uneven Time Intervals

A common error when using Excel is to draw a bar chart or line chart for a time series with time intervals of varying length. The time periods between the evenly spaced tick marks on the *x* axis are inconsistent. The top plot in Figure 9.2 hangs in a U.S. museum; only the *y* axis label has been changed to protect privacy. Data are available for irregularly spaced intervals. The default in Excel for bar charts and line charts are labels rather than numbers on the horizontal axis (i.e., Excel assumes that the data are categorical rather than quantitative). The bottom chart uses an *xy* plot or scatterplot which assumes that both variables are quantitative. The *xy* plot spaces the years correctly. It is important to choose your chart type according to your data rather than the look of the finished chart when using Excel.

Fig. 9.3 State Areas: Dot Chart from Excel

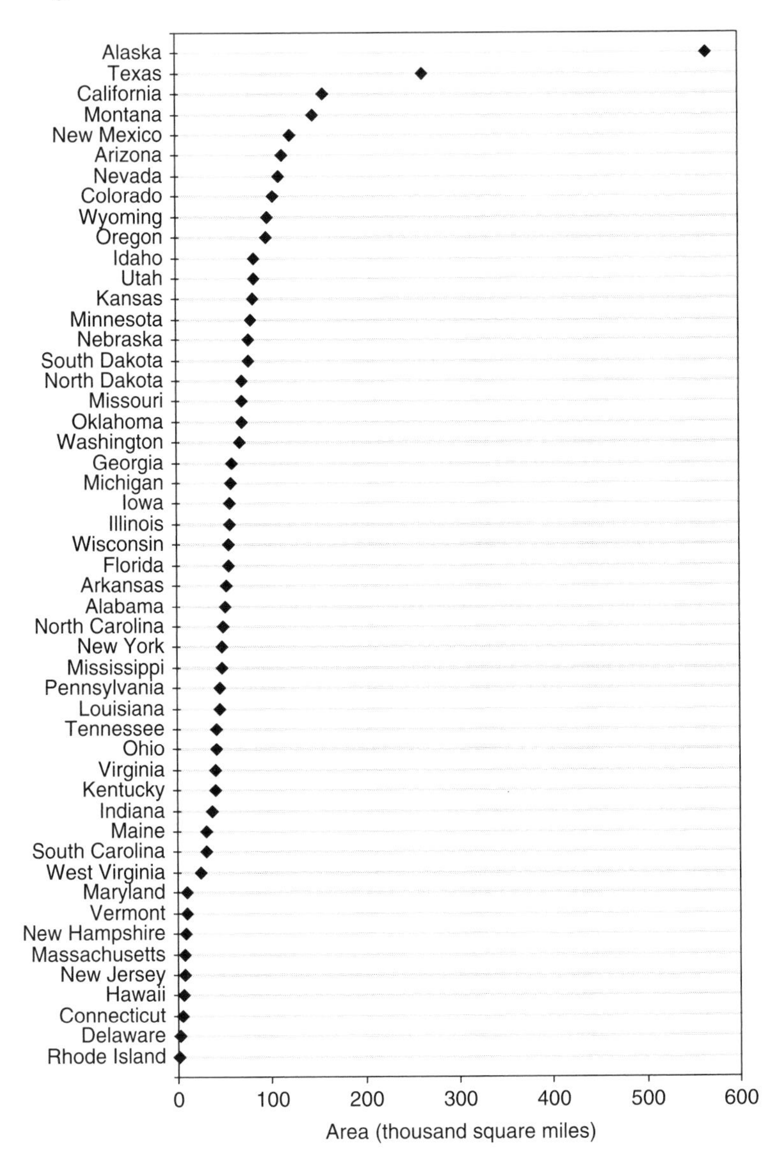

9.3.3 Dot Charts from Excel

I use dot plots frequently. The dot chart of state areas in Figure 4.3 was drawn using S-Plus. Figure 9.3 was drawn using Excel. Excel menus do not offer charts that look like this. To make one, use an *xy* plot with the areas as the *x* variable and the numbers 1 through 50 as the *y* variable. Then convert the *y* labels (the numbers from 1 through 50) to labels using a macro written for this purpose by Kenneth Klein. Ken has agreed to let readers of this book use his macro. It is available for downloading at

ftp://ftp.wiley.com/public/sci_tech_med/graphs/

Instructions for its use are given in the spreadsheet containing the macro.

Fig. 9.4 Marketing Data: Data Labels with Excel

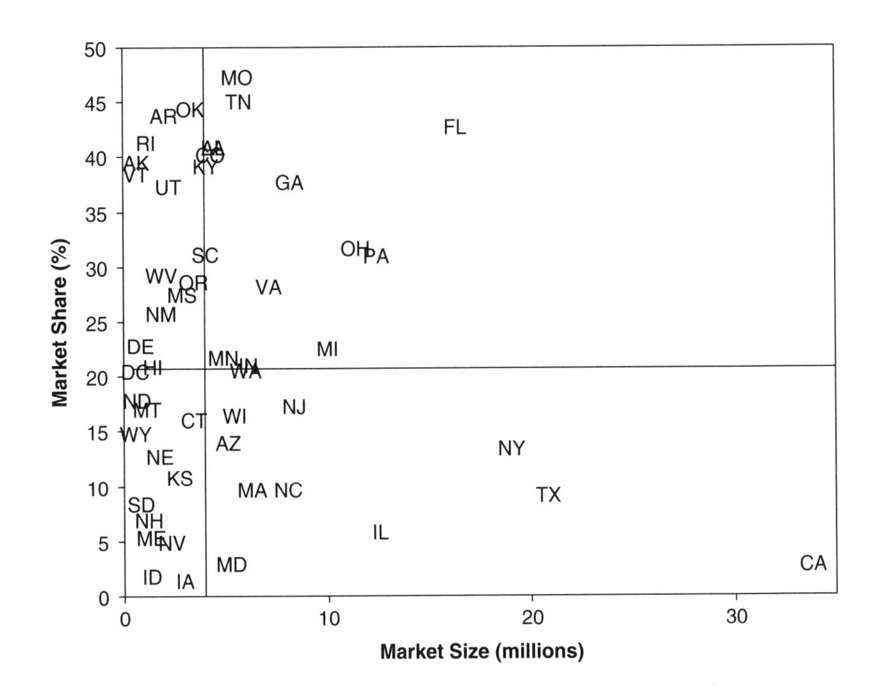

9.3.4 Data Labels with Excel

It is often useful to label data points with identifying information, such as the two-letter state abbreviations used in Figure 9.4. It is easy to label points with their numerical values in Excel, but is not as easy to do with text. I use a free utility called *xy labeler* available for download at

http://www.appspro.com/utilities/Labeler.asp

to add other labels. This utility allows you to add labels automatically, move labels, or add some labels manually.

Figure 9.4 duplicates Figure 6.12, drawn using S-Plus, in Excel. The horizontal line indicates the median value of market share, and the vertical line indicates the median value of market size. To reduce clutter, I positioned the labels on top of the data points and then removed the data points.

We have just seen that Excel is capable of producing some useful graph forms that don't appear on its menus. Some require add-in utilities that can be downloaded from the Web. Perhaps readers who are adept at using Excel will write additional add-ins or macros and share them with others.[2] We also learned that some Excel options create confusing and misleading graphs. A review of Chapter 2 will remind you of options in Excel that you should avoid. You can improve your graphs dramatically by changing defaults and choosing options wisely.

[2] If you do and want me to link to it on my Web page, contact me at *naomi@nbr-graphs.com*.

SUMMARY

Charts are produced today in a surprising variety of software environments. The commands of the S language offer unlimited options but require an understanding of computer programming. The graphical user interface of S-Plus captures many of the advantages of S in a more user-friendly environment. Excel, the popular spreadsheet program, can produce acceptable graphs if options are chosen wisely and defaults are overruled. Finally, some drawing programs (e.g., Illustrator) offer superior design options and fine control over graphical elements.

10 Questions and Answers

A seminar usually ends with questions and answers, and this has been a seminar in book form. I have frequently been asked the following questions:

1. When should I use a table, and when should I use a graph?
2. Should I use different graphs for presentations and for written reports?
3. How do graphs for data analysis and graphs for communication differ?
4. What should I use instead of pie charts?

Creating More Effective Graphs, by Naomi B. Robbins
ISBN 0-471-27402-X Copyright © 2005 John Wiley & Sons, Inc.

5. What if I just want to give an impression of the direction of the data? Then may I use three-dimensional charts?
6. I use three-dimensional charts but I include data labels. That's OK, isn't it?
7. I want my graphs to attract the reader's attention. How should I decorate them?
8. Why do you think we see so many bad graphs?
9. When should I use each type of graph?

1. When should I use a table, and when should I use a graph?

Graphs are for the forest and tables are for the trees. Graphs give you the big picture and show you the trends; tables give you the details. Graphs may be used with most media: paper, projection screen, or computer screen. Large tables, on the other hand, do not work well on projection screens.[1]

The U.S. Bureau of Labor Statistics provides unemployment rate from 1948 until the present. The table in Figure 10.1 shows the data from 1948 to 1962; the graphs in Figure 10.2 cover the period from 1948 to 2000. The first thing we note is that

[1] An example is given in Section 8.5.

the graph takes much less room when there is a large amount of data. Questions such as "When was the unemployment rate the highest?" or "How many cycles have there been?" are much easier to answer from the graph. However, a researcher needing the unemployment rate for a specific month to input into a model would do better with the table.

My preference is to show both. However, showing both takes more paper or more time in a presentation. Therefore, the choice depends on your purpose and how the information will be used.

Fig. 10.1 BLS Unemployment Data: Table

Year	Jan	Feb	Mar	Apr	May	Jun	Jul	Aug	Sep	Oct	Nov	Dec
1948	3.4	3.8	4.0	3.9	3.5	3.6	3.6	3.9	3.8	3.7	3.8	4.0
1949	4.3	4.7	5.0	5.3	6.1	6.2	6.7	6.8	6.6	7.9	6.4	6.6
1950	6.5	6.4	6.3	5.8	5.5	5.4	5.0	4.5	4.4	4.2	4.2	4.3
1951	3.7	3.4	3.4	3.1	3.0	3.2	3.1	3.1	3.3	3.5	3.5	3.1
1952	3.2	3.1	2.9	2.9	3.0	3.0	3.2	3.4	3.1	3.0	2.8	2.7
1953	2.9	2.6	2.6	2.7	2.5	2.5	2.6	2.7	2.9	3.1	3.5	4.5
1954	4.9	5.2	5.7	5.9	5.9	5.6	5.8	6.0	6.1	5.7	5.3	5.0
1955	4.9	4.7	4.6	4.7	4.3	4.2	4.0	4.2	4.1	4.3	4.2	4.2
1956	4.0	3.9	4.2	4.0	4.3	4.3	4.4	4.1	3.9	3.9	4.3	4.2
1957	4.2	3.9	3.7	3.9	4.1	4.3	4.2	4.1	4.4	4.5	5.1	5.2
1958	5.8	6.4	6.7	7.4	7.4	7.3	7.5	7.4	7.1	6.7	6.2	6.2
1959	6.0	5.9	5.6	5.2	5.1	5.0	5.1	5.2	5.5	5.7	5.8	5.3
1960	5.2	4.8	5.4	5.2	5.1	5.4	5.5	5.6	5.5	6.1	6.1	6.6
1961	6.6	6.9	6.9	7.0	7.1	6.9	7.0	6.6	6.7	6.5	6.1	6.0
1962	5.8	5.5	5.6	5.6	5.5	5.5	5.4	5.7	5.6	5.4	5.7	5.5

Fig. 10.2 BLS Unemployment Data: Graph

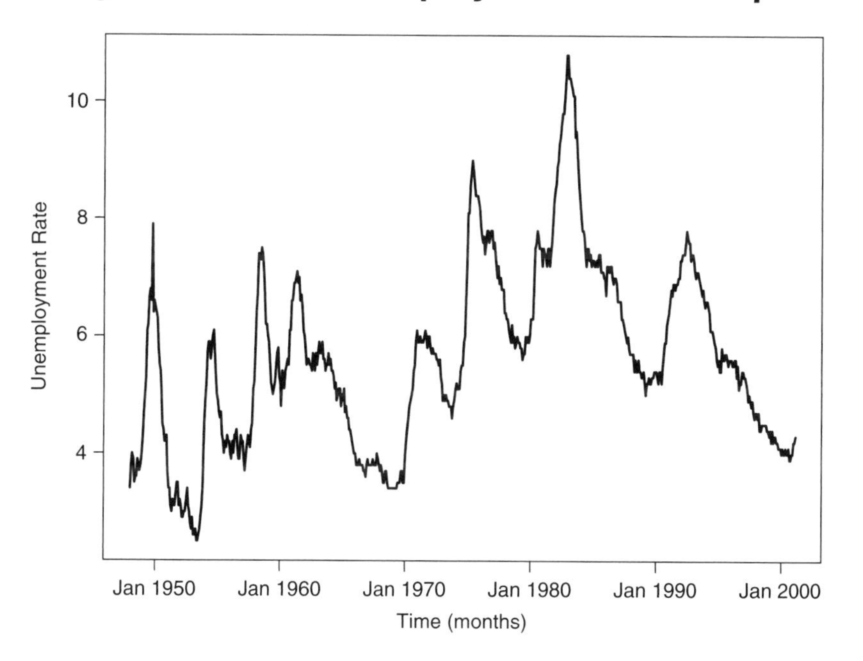

2. Should I use different graphs for presentations and for written reports?

The principles of drawing effective graphs are the same no matter what the medium: strive for clarity and conciseness. However, since a reader may spend more time studying a written report than is possible during a presentation, more detail can be included. Figures that are projected must have large type fonts and be legible from a distance. I'm often asked which graphs are suitable for presentations. It depends on the audience. A figure that is commonplace for one group may appear complicated and difficult to understand for another. For example, scatterplot matrices may be used regularly at a statistical conference or a research laboratory meeting, but may be difficult for a general audience to understand.

3. How do graphs for data analysis and graphs for communication differ?

Kosslyn (1994) and other authors distinguish between graph forms suitable for presentation and those to be used for data analysis. I don't find this distinction very useful. Many of the common graph forms for presentation have perceptual problems that prevent you and the reader from gaining an early understanding of the data. On the flip side, a late step in research and data analysis is the communication of results. If you find that one of your analysis graphs makes a powerful point, why abandon it for a simpler "presentation form" that cannot communicate your idea as effectively? However, the analysis graph does not need to be presentation quality. When you're ready to show your graph to others, it's time to adjust the font size; add clear titles, axis labels, and legends; and attend to other details.

4. What should I use instead of pie charts?

We have seen in early chapters that a pie chart does not communicate very effectively. We noted that dot plots and other charts that require judgments of position along a common scale are better for getting quantitative information across reliably. But we also noted that there may be other reasons to draw graphs, such as to make a page more inviting (see pages 6 and 7). Many financial reports use pie charts with three or four wedges to decorate the page. If this is your purpose, stick with a two-dimensional chart since a three-dimensional chart distorts your data. Adding data labels to a pie chart will make the information clearer.[2] But realize, of course, that you are communicating through the text in the data labels rather than through the chart itself.

[2] Question 6 deals with data labels on bar charts.

Fig. 10.3 Presidential Approval Data: Waffle Chart

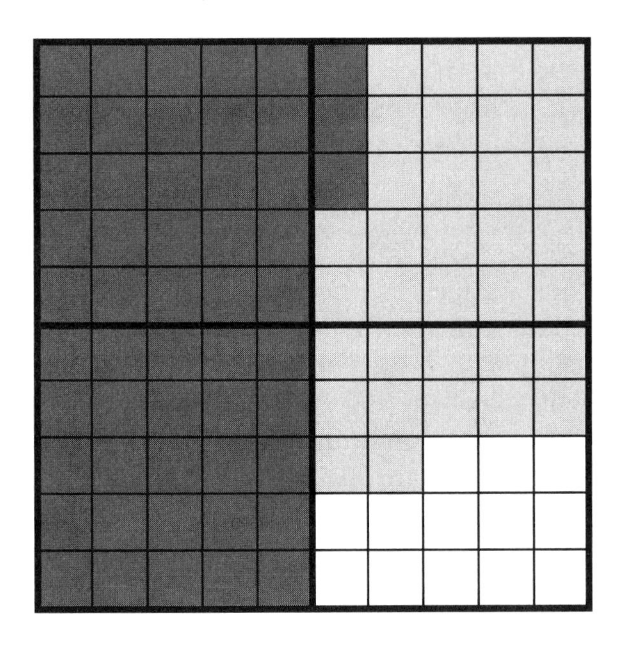

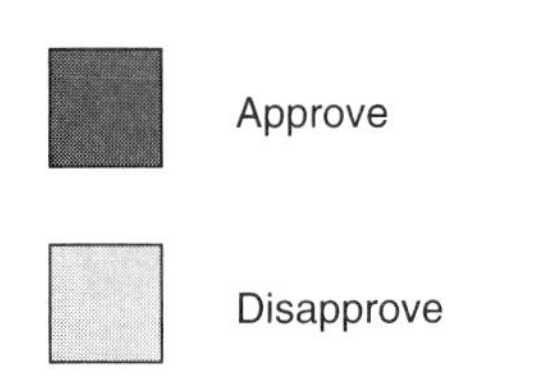

One alternative to a pie chart is a chart that I first saw in the *New York Times* in June 2001 to communicate presidential approval ratings. The designer showed a series of charts showing approval and disapproval rates for how the president was handling his job, the environment, and other categories. Figure 10.3 reproduces their figure on how the president was handling his job. Although one of these charts is overkill for two numbers, some find it easier to compare multiple waffle charts than multiple pie charts.

To draw the chart I created a table in Word with 10 rows and 10 columns, then shaded 53 squares with one fill color to show that 53% approve, and 34 squares with a different fill color for the 34% who disapprove.

Fig. 10.4 Presidential Approval Data: Which Do You Prefer?

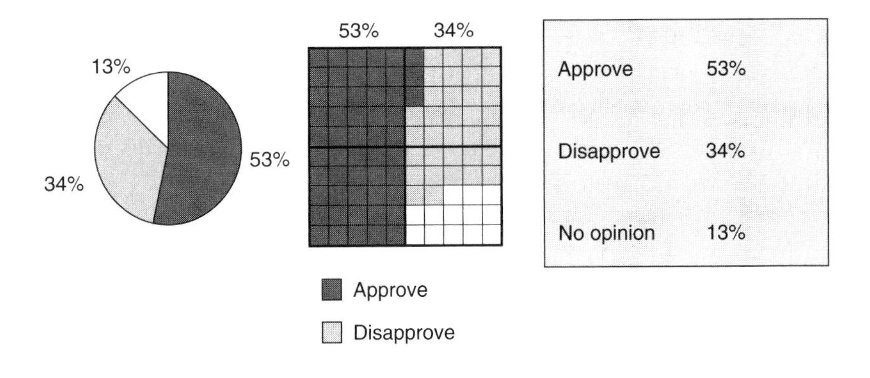

Whatever your choice, you are not alone. Some people find a circle to be pleasing, others prefer the fact that they can determine the value from the waffle chart, and still others prefer the text box. When there is only one set of numbers my preference is for the text box. A sentence is sufficient for the small amount of information here, but a sentence might not be noticed.

Fig. 10.5 Total Sales: Three-Dimensional Bar Chart

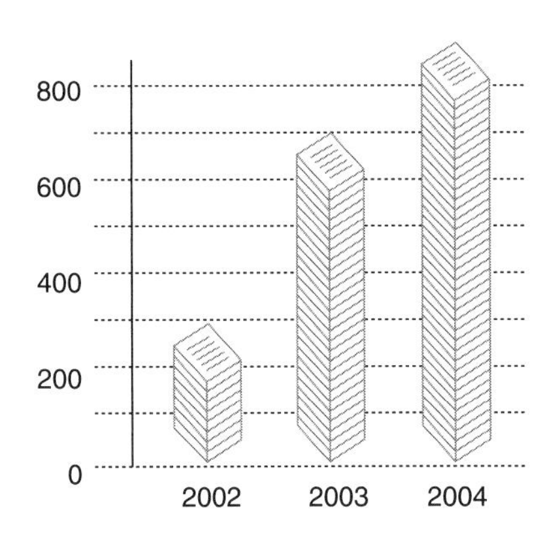

5. What if I just want to give an impression of the direction of the data? Then may I use three-dimensional charts?

My answer is "no."

Often, charts are used to show trends: for example, to show that sales have been increasing over time. The designer of the chart wants to make it clear that there has been an improvement and is not as interested in communicating the actual numbers. Some graph designers who think that three-dimensional bar charts are attractive ask if they can use them in that case.

The problem is that although the graph designer may only be interested in showing the trend, the reader might be very interested in knowing the actual number. The spread of the top of the 2002 bar in Figure 10.5 goes from around 170 to 300. That is too large a range to help the reader determine the actual number. Writers don't use vague words because the reader just needs an impression of what is intended. Let's show the same respect for numbers.

Fig. 10.6 Data Labels on Three-Dimensional Bars

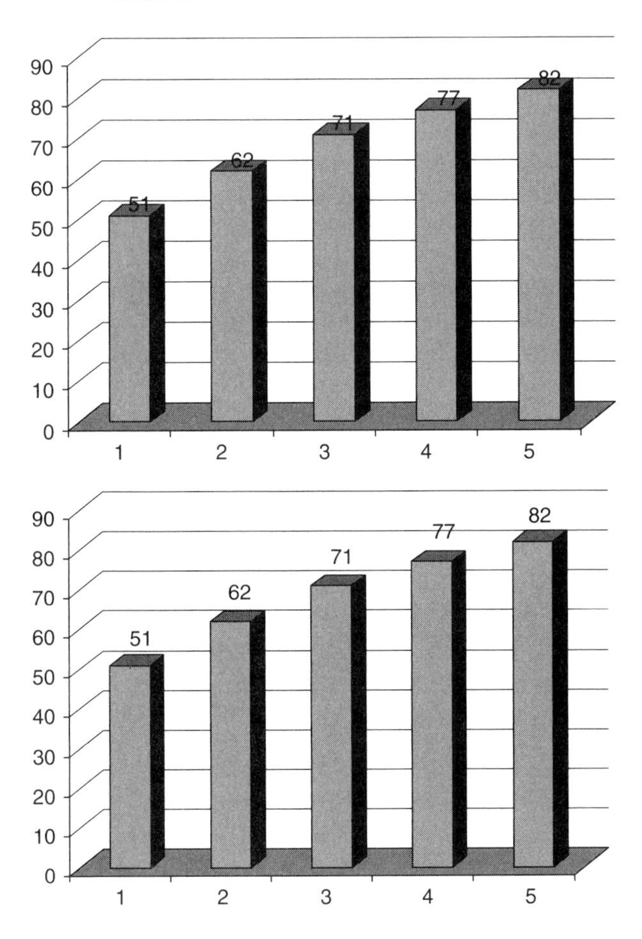

6. I use three-dimensional charts but I include data labels. That's OK, isn't it?

Again, my answer is "no."

The top chart in Figure 10.6 includes data labels placed in the default position for this type of chart in Excel. They confuse more than they clarify. Look at the middle bar. It is labeled 71 but appears visually to be below 70. I attended a presentation where many of the charts looked like this one. I heard a buzzing in the audience: People were whispering to their neighbors that the speaker was trying to mislead them since the bars did not match the numbers. Moving the labels to make them more legible, as we see in the bottom chart, solves one problem but accentuates the problem of the bars appearing to be lower than the value they represent. Bottom line: Do not use these charts with or without labels.

7. I want my graphs to attract the reader's attention. How should I decorate them?

This is a controversial topic. My position is that there are times when changing the mood of a graph is appropriate, but this should be done outside the scale-line rectangle without interfering with the data.

Books about graphs written by statisticians often say or imply that "a graph that calls attention to itself pictorially is almost surely a failure" (Wainer, 1997, p. 11). On the other hand, books on graphs for journalists and graphic designers suggest ways to catch a reader's attention and motivate the reader to examine the data by adding ornamentation to the graph.

In addition to the author's purpose in creating the figure, let's consider the reader's purpose in reading the figure. First, suppose that you are an executive who needs to make an important business decision. You are presented with data in graphical form to provide the background needed to make that decision. Certainly, the data are of interest without any adornment. You want the facts and you want them to be clear and accurate. Now, suppose that you are the same executive, sitting in the waiting room of a doctor's office, annoyed that you have to wait. You pick up a magazine and skim through it looking for an article to read to kill time. You read an article that attracts your attention. It might be the title, the page layout, or perhaps a graph that grabs your attention. The figure that satisfied your needs for the business decision might go unread in the doctor's office. However, catching readers' attention never justifies distorting data or misleading readers. In this section we learn how to strike a balance between the extreme positions of the prototypical statistician and graphic artist.

Fig. 10.7 Computer Sales: Framing the Graph

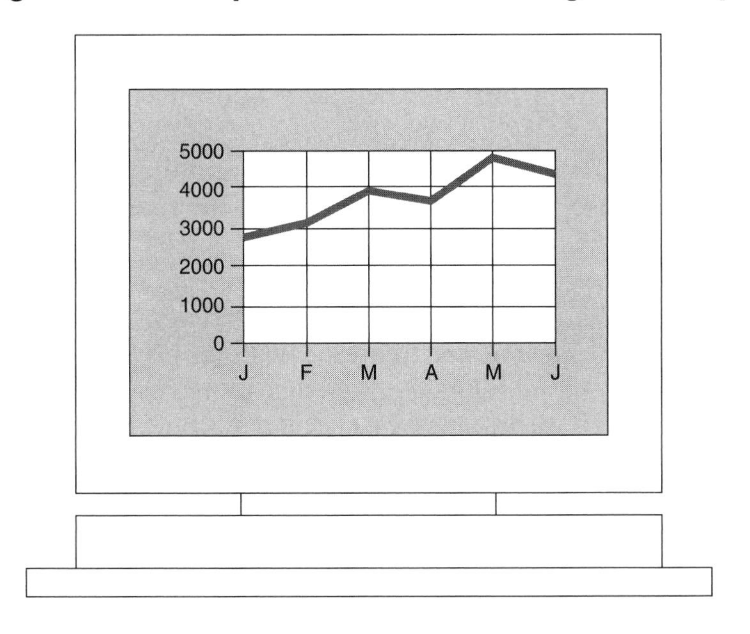

The challenge, then, is to embellish a figure without violating the principles we just learned. One way to do this is to keep the decorations outside the area containing the data. Kostelnick (1998) discusses conflicting philosophies of data display. He frames the computer sales graph with a computer screen to grab attention and announce the subject matter. "These kinds of artful displays can radically alter visual tone," he claims, "making it more informal and conversational, maybe even playful—though the use of such artistry horrifies informational purists like Tufte (1983), who dismisses it as mere 'chartjunk'."

Chartjunk is a term coined by Tufte to refer to nonessential ink used in the interior decoration of graphics. It includes such things as moiré effects, prominent grid lines, and redundant representations of data. I do not consider the computer frame in Figure 10.7 to be chartjunk; it is not part of the chart.

Fig. 10.8 Kostelnick's Informal Tone

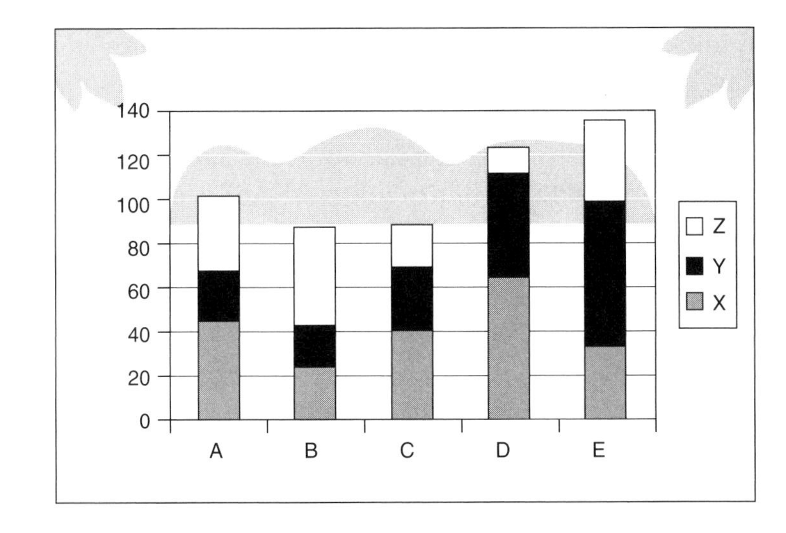

I have seen some excellent graphs in the *New York Times* that attracted attention through framing. The charts are uncluttered, with the data standing out clearly. The frame was the part that called attention to itself.

Figure 10.8 shows a display appropriate for vacationers. The gray that looks like clouds in the background is a profile of an island to make the figure more informal and inviting. I consider that profile to be chartjunk.

Fig. 10.9 Informal Tone without Graph Clutter

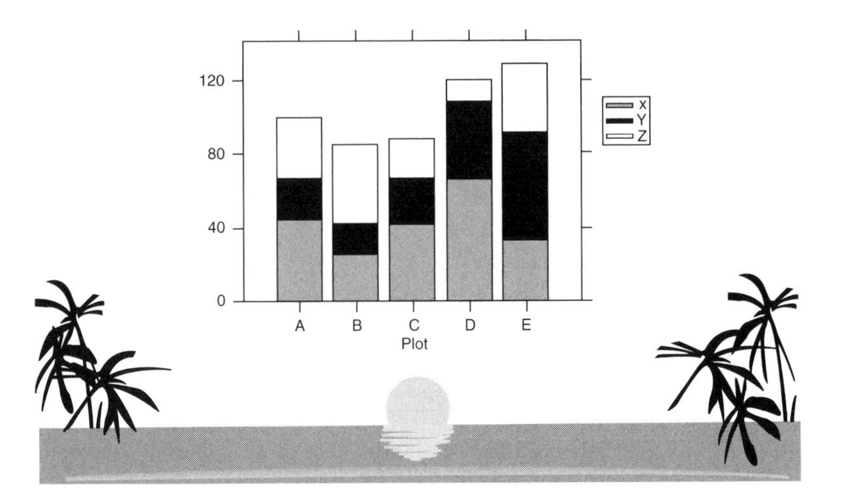

Figure 10.9 shows an alternative way to achieve a similar mood without cluttering the data. Although I generally don't use stacked bar charts because of their perceptual problems, I wanted to keep everything but the decoration reasonably constant. I certainly agree that differing audiences require different presentations. I agree that at times it is appropriate to change a mood or attract attention using decorations. But that does not mean that we need to violate the principles of creating clear, accurate graphs.

Fig. 10.10 Adding Glitz with Labels

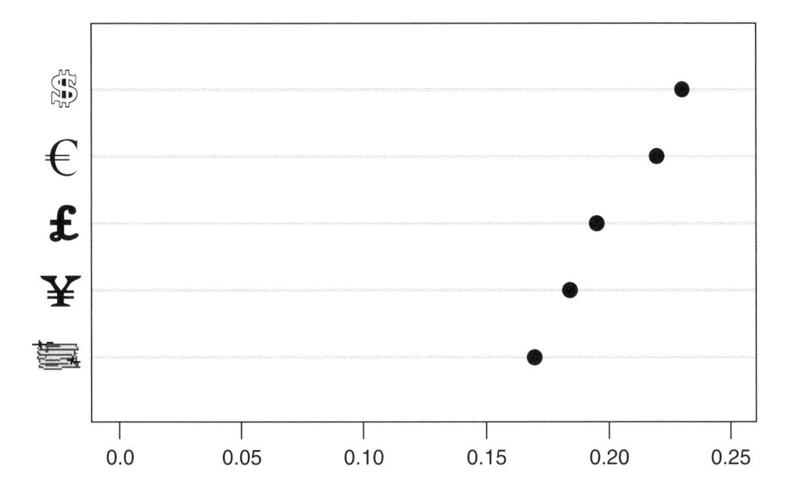

Suppose that the categories of Figure 10.10 represent different forms of payment. These pictorial labels attract attention without distracting from the data. They have the added advantage of not being language dependent, so that the figure can be used for an international audience.

To summarize, a graph should be appropriate for the audience and the situation in which it is shown. This can be done without adding clutter inside the scale-line rectangle. The data can still stand out. I have seen some very effective graphs in corporate annual reports that attracted my attention due to their originality. Yet they had scales, were drawn to scale, and were concise, accurate representations of the data.

8. Why do you think we see so many bad graphs?

I'm often asked: "Are most of the problems we saw in previous chapters due to the graph designer or using the wrong software or using the software incorrectly?"

Many factors contribute to the quality of the graphs we regularly see:

1. We learn from what we see, and unfortunately, we see lots of three-dimensional charts and other charts with perceptual problems.
2. Some graph designers mislead intentionally to give the appearance of better performance than is actually the case (e.g., Figure 6.24).
3. Many users settle for the default chart their software produces even though changes would make the data stand out better.
4. Some artists who have little appreciation for numbers draw graphs without understanding the need to draw to scale.

5. Sometimes the software allows the user to make unreasonable choices (e.g., Figure 4.7).
6. Sometimes the software itself makes unreasonable choices (e.g., double-y axes in Excel; see Section 7.5).
7. Designers may be more concerned with impressing readers with their technological capabilities or artistic abilities than with communicating clearly.
8. We receive much more training in communicating with words than in communicating with numbers.
9. Many of the problems we have seen were caused by carelessness and lack of proofreading.

It is worth noting that graphs have improved dramatically in the last 20 years, thanks to the excellent books by Cleveland (1985) and Tufte (1983), which have encouraged large numbers of readers to produce excellent, clear graphs.

9. When do I use each type of graph?

Your question implies that there is a finite list of graph types and concrete rules for using each type. Even if that were true, this book makes no pretension of being complete. Rather, it presents a few useful graph forms. But the number of possible graph presentations is limited only by your imagination. The graph designer should think about the data and how those data are best presented. The radar graph in Chapter 8 consisted of a separate set of questions in each of four categories. Trellis displays repeat the same information in each panel. The custom trellis plot shown in Figure 8.9 was motivated by trellis principles but was customized for the data it shows.

Similarly, it is meaningless to assign graph types to fields of application. Recall that data may be quantitative or categorical (labels such as male or female). A bar chart is used to compare the numerical values for categorical data such as literacy rate by country (category). It doesn't matter whether we are comparing the number of home runs for the top baseball hitters, the heights of the tallest mountains, or the sales of a drug for each month of a year. The issues are the same: A bar chart without zero misleads, and bar charts get cluttered quickly. All of these examples have large values, so including zero will not show the comparisons of interest clearly. The type of graph to use depends on the nature of the data you are working with, not on your profession!

Even graphs created specifically for one industry find application elsewhere. The high-low charts created to display stock market data are applicable whenever we want to display a range of values over time. They are useful for displaying weather charts showing temperature highs and lows. That said, I will provide some guidance about when to use the chart types we mentioned, but in terms of the data characteristics, not the field of application.

Bar charts: Use two-dimensional bar charts as a second choice to dot plots for showing comparisons among categorical data.

Box plots: Use for showing the distribution of one data set and for comparing distributions of multiple data sets.

Dot plots: Extremely effective for showing categorical data. Dot plots can be used for any situation for which a bar chart is commonly used.

Grouped bar chart: Use for comparisons among several sets of categorical data. Multiway dot charts are more effective.

Grouped or labeled scatterplot: Use when there are two quantitative variables and a categorical variable (e.g., heights, weights, and gender) to see the relationship of the two quantitative variables for each group and to compare the groups. Different plotting symbols or colors are used for the different groups.

High–low–open–close plot: Use to display financial data.

Histograms: Use to show the distribution of a single variable. Not useful for comparing more than one variable.

Line graph: Use to connect points in a graph with either two quantitative variables or one quantitative and one ordered categorical variable. Do not use with a nonordered categorical variable.

Linked micromaps: Use for geographically referenced data.

Month plot: Use to show the behavior of subseries, such as the days of the week or the months of the year.

Mosaic plots: Use when there is one quantitative and several categorical variables to show the counts of cross-classified data.

Parallel coordinate plots: Use for displaying multivariate data sets.

Pie charts: If at all, use only to show parts of a whole (data that are, or could be, expressed in percentage form).

Scatterplot: Use for plotting the relationship between two quantitative variables.

Scatterplot matrix: Use for studying the relationship between all pairs of variables when there are more than two quantitative variables.

Stacked bar chart: Do not use, because of their perceptual problems.

Strip plots: Use to show the distribution of one-dimensional quantitative variables. Unlike box plots and histograms, strip plots show all the data.

Three-dimensional pie or bar charts: Do not use because they are difficult to read and mislead the reader.

Time-series plot: Use when one of the variables is a unit of time.

Trellis displays: Use to look at two or three variables for all combinations of remaining variables.

APPENDIX A

Checklist of Possible Graph Defects

CAN THE READER CLEARLY SEE THE GRAPHICAL ELEMENTS?

Do the data stand out? Are there superfluous elements? (159)

Are all graphical elements visually prominent? (163)

Are overlapping plotting symbols visually distinguishable? (165)

Can superposed data sets be readily visually assembled? (167)

Is the interior of the scale-line rectangle cluttered? (175)

Do data labels interfere with the quantitative data or clutter the graph? (175)

Is the data rectangle within the scale-line rectangle? (179)

Do tick marks interfere with the data? (179)

Do tick mark labels interfere with the data? (179)

The numbers in parentheses indicate the page numbers on which I discuss the items.

Creating More Effective Graphs, by Naomi B. Robbins
ISBN 0-471-27402-X Copyright © 2005 John Wiley & Sons, Inc.

Are axis labels legible? (180)

Are there too many tick marks? (183)

Are there too many tick mark labels? (183)

Do the grid lines interfere with the data? (185)

Are there notes or keys inside the scale-line rectangle? (189)

Will visual clarity be preserved under reduction and reproduction? (191)

CAN THE READER CLEARLY UNDERSTAND THE GRAPH?

Are the data drawn to scale? (197)

Is there an informative title? (157)

Is area or volume used to show changes in one dimension? (203)

Are there too many dimensions in the graph (more than in the data)? (25-27)

Are common baselines used wherever possible? (207)

Are all labels associated with the correct graphical elements? (195, 215)

Is the reader required to make calculations? (216)

Are groups of charts drawn consistently? (221)

ARE THE SCALES WELL CHOSEN AND LABELED?

Is zero included for all bar graphs? (239)

Are there any unnecessary scale breaks? (257)

Is there a forceful indication of a scale break? (257)

Are there numerical values on two sides of a scale break that are connected? (261)

Does the aspect ratio allow the reader to see variations in the data? (269)

Are scales included for all axes? (277)

Are the scales labeled? (279)

Are tick marks at sensible values? (281)

Do the axes increase in the conventional direction? (283)

Does the data rectangle fill as much of the scale-line rectangle as possible? (285)

Are uneven time intervals handled correctly? (287, 335)

Are the scales appropriate when different panels are compared? (289)

APPENDIX B

List of Figures with Sources

Figure Number and Title	Page	Source of Data	Software Used/ Source of Figure
1.1 Similar Pie Wedges	2		S-Plus; inspired by Cleveland (1985)
1.2 Similar Pie Wedges: Dot Plot	4		S-Plus; inspired by Cleveland (1985)
1.3 Similar Pie Wedges: Table	8		Word
2.1 Structured Data Set	12		S-Plus; inspired by Cleveland (1994)
2.2 Structured Data Set: Dot Plot	14		S-Plus; inspired by Cleveland (1994)
2.3 Three-Dimensional Pie Data	18		Excel
2.4 Three-Dimensional Pie Data: Two-Dimensional Bar Chart	20		Excel
2.5 Three-Dimensional Pie Data: Two-Dimensional Pie Chart	22		Excel
2.6 Three-Dimensional Bar Data	22		Excel
2.7 Three-Dimensional Bar Data: Two-Dimensional Bar Chart	24		Excel

Creating More Effective Graphs, by Naomi B. Robbins
ISBN 0-471-27402-X Copyright © 2005 John Wiley & Sons, Inc.

Figure Number and Title	Page	Source of Data	Software Used/ Source of Figure
2.8 Three-Dimensional Bar Data: Excel	24		Excel
2.9 Three-Dimensional Bar Data: PowerPoint	26		PowerPoint
2.10 Three-Dimensional Bar Data: Presentations and Charts	26		Presentations & Charts
2.11 Energy Data	28	U.S. Dept. Energy (1986)	S-Plus
2.12 Energy Data: All Other OECD	30	U.S. Dept. Energy (1986)	Excel
2.13 Energy Data: Grouped Bar Chart	32	U.S. Dept. Energy (1986)	Excel
2.14 Playfair's Balance-of-Trade Data	34	S-Plus 6 for Windows User's Guide. Copyright © 2003 Insightful Corporation, Seattle, WA.	S-Plus; from Cleveland (1994)
2.15 Playfair's Balance-of-Trade Data: Imports Minus Export	36	S-Plus Data Set	S-Plus; from Cleveland (1994)
2.16 Difference between Curves	38		S-Plus
2.17 Ownership of Government Securities	40	Board of Governors of the Federal Reserve System (U.S.) (1989); Historical Chart Book, Washington, DC.	
2.18 Playfair's Population of Cities	42	S-Plus Data Set	S-Plus; from Cleveland (1993)
2.19 Population of Cities: Dot Plot	44	S-Plus Data Set	S-Plus; from Cleveland (1993)
3.1 Angle Judgments	48		S-Plus
3.2 Area and Volume Judgments	50		S-Plus
3.3 Color Hue, Saturation, and Density	52	Random numbers	S-Plus

Figure Number and Title	Page	Source of Data	Software Used/ Source of Figure
5.9 MBA Data: Another Variable Added	132	Monterey Bay Aquarium	S-Plus with *http://www.stat.yale.edu/~jay/JCGS/mosaic.code*
5.10 MBA Data: Mosaic Plot	134	Monterey Bay Aquarium	S-Plus with *http://www.stat.yale.edu/~jay/JCGS/mosaic.code*
5.11 Soybean Data: Linked Micromaps	136	*http://www.nass.usda.gov/research/gmsoyyap.htm*	
Choropleth Map	139	Carr (1993)	
5.12 Iris Data: Parallel Coordinate Plots	140	S-Plus Data Set	S-Plus
5.13 Reading a Parallel Coordinate Plots	142		S-Plus
5.14 Nightingale Data	144	*http://www.uh.edu/engines/epi1712.htm*	
5.15 Nightingale Data: Trellis Plot	148	Provided by Hugh Small	S-Plus
5.16 DJIA: High-Low-Open-Close Plots	150	Dow Jones Industrial Average	S-Plus
6.1 Terminology	156	St. Louis Science Center	S-Plus
6.2 Association of Research Libraries Data	158	ARL Notebook 2002 LibQUAL + (tm) Results (Washington, DC: Association of Research Libraries, 2002). 4.1.2: Respondents by Discipline for 4-Year Institution—All User Groups (Includes Library Staff)[table], p. 26. Retrieved January 20, 2003 from <*http://www.libqual.org/documents/admin/ARLNotebook111.pdf*> For more information on the ARL Statistics and Measurement Program, see <*http://www.arl.org/stats/*>	
6.3 ARL Data: Data Stand Out Better	160	Association of Research Libraries (2002)	S-Plus
6.4 Data That Are Difficult to See	162		S-Plus
6.5 Museum Exhibitions: Overlapping Data	164	Beverly Serrell	S-Plus

Figure Number and Title	Page	Source of Data	Software Used/ Source of Figure
6.23 Police Officer Data to Scale	200	State of New Jersey, Division of State Police, Uniform Crime Reporting Unit (1997)	Excel
6.24 Earnings per Share Data	202		Top: Linda Clark using S-Plus, inspired by graphs in annual reports; bottom: Excel
6.25 America Online Market Value	204	Wurman (1999)	Top: Wurman (1999); bottom: Becker Chart
6.26 Annual Report Data	206		Linda Clark using S-Plus; inspired by graphs in annual reports
6.27 Immigration Data	208	Wurman (1999)	
6.28 Immigration Data with Horizontal Baseline	210	Wurman (1999)	Drawn by Marc Tracey using Illustrator. Used with his permission.
6.29 Survey Results Data	212		Excel; inspired by graphs in newsletters
6.30 Health Insurance Data	214	Wurman (1999)	Top: Wurman (1999); bottom: Excel
6.31 Source of Funding Data	218		Top: Linda Clark using S-Plus; inspired by graphs in newsletters; bottom: Excel
6.32 Population of Cities	220	U.S. Census	S-Plus
6.33 Population of Cities: Conflicting Principles	222	U.S. Census	S-Plus
7.1 Sunspot Data: Aspect Ratio 1	228	S-Plus Data Set	S-Plus; from Cleveland (1994)
7.2 Annual Report: Aspect Ratio 2	230		S-Plus; top inspired by graphs in annual reports

Figure Number and Title	Page	Source of Data	Software Used/ Source of Figure
7.3 Carbon Dioxide Data	234	S-Plus Data Set	S-Plus; from Cleveland (1994)
7.4 Carbon Dioxide: Aspect Ratio Changed	236	S-Plus Data Set	S-Plus
7.5 Annual Report Bar Chart: No Zero	238		Excel; top inspired by graphs in annual reports
7.6 Logarithmic Scales for Percent Change	242		S-Plus
7.7 Sales Data	244	Private client	S-Plus
7.8 Sales Data: Logarithmic Scale	246	Private client	S-Plus
7.9 Sales Data: Ordered Data with Logarithmic Scale	248	Private client	S-Plus
7.10 Police Data: Logarithmic Scale with Bars	250	State of New Jersey, Division of State Police, Uniform Crime Reporting Unit (1997)	
7.11 Police Data: Scale Break	252	State of New Jersey, Division of State Police, Uniform Crime Reporting Unit (1994)	
7.12 Police Data: Simple View	254	State of New Jersey, Division of State Police, Uniform Crime Reporting Unit (1997)	Excel
7.13 Sales Data: Full Scale Break	256	Private client	S-Plus
7.14 Sales Data: Avoiding a Scale Break	258	Private client	S-Plus
7.15 Connecting Sides of a Scale Break	260		Linda Clark using S-Plus; inspired by graphs in annual reports
7.16 Temperature Data: Two Scales	262	Sutcliffe (1996)	S-Plus
7.17 Income Data: Gender Gap at All Ages	264	U.S. Bureau of Labor Statistics	S-Plus
7.18 Income Data: Gender Gap Depends on Age	266	U.S. Bureau of Labor Statistics	S-Plus
7.19 Blood-Level Data	268	Private communication	Linda Clark using S-Plus

Figure Number and Title	Page	Source of Data	Software Used/ Source of Figure
7.20 Blood-Level Data: Variation in Ratio	270	Private communication	Linda Clark using S-Plus
7.21 Blood-Level Data: Multipanel Plot	272	Private communication	Linda Clark using S-Plus
7.22 Blood-Level Data: Scatterplot Matrix	274	Private communication	R
7.23 Annual Report: No *Y* Scale	276		S-Plus; top inspired by graphs in annual reports
7.24 Prime Lending Rate Data: No *X* Scale	278	Copyright © 2003 - 2004 Spydercube, Inc. Used with their permission.	
7.25 Annual Report Data: Tick Marks at Hard-to-Judge Values	280		Excel; inspired by graphs in annual reports
7.26 IRS Data: Departing from Convention	282	*http://www.gao.gov*	
7.27 Invoice Data: Filling the Scale-Line Rectangle	284	Texas State Auditor's Office (1995)	
7.28 Body Mass Data: Unequal Intervals	286	Calle et al. (1999). Copyright © 1999 Massachusetts Medical Society. All rights reserved.	
7.29 Body Mass Data: Comparing Panels	288	Calle et al. (1999). Copyright © 1999 Massachusetts Medical Society. All rights reserved.	
7.30 Body Mass Data: Corrected	290	Copyright © 1999 Edith Flaster. Used with her permission.	
8.1 Gold Price Data	294	*http://www.gold-eagle.com/analysis/london_gold.html*	Excel; inspired by graphs on Web
8.2 Gold Price Data: High - Low Chart	296	*http://www.gold-eagle.com/analysis/london_gold.html*	S-Plus
8.3 Task Data	298	*Dynamics in Document Design: Creating Texts for Readers* by Karen A. Schriver. Copyright © 1997 by John Wiley & Sons, Inc. All rights reserved. Reproduced here by permission of the publisher.	
8.4 Task Data Unified	300	Schriver (1997)	S-Plus
8.5 Task Data: Improvement	302	Schriver (1997)	S-Plus

Figure Number and Title	Page	Source of Data	Software Used/ Source of Figure
10.2 BLS Unemployment Data: Graph	347	U.S. Bureau of Labor Statistics	S-Plus
10.3 Presidential Approval Data: Waffle Chart	352	*New York Times*, June 21, 2002	Word
10.4 Presidential Approval Data: Which do you prefer?	354–355	*New York Times*, June 21, 2002	Left: Excel; middle: Word
10.5 Total Sales: Three-Dimensional Bar Chart	356		Paint; inspired by ads in junk mail
10.6 Data Labels on Three-Dimensional Bars	358		Excel; inspired by presentations
10.7 Computer Sales: Framing the Graph	362	Written by Charles Kostelnick and reprinted with permission from the Society for Technical Communication, Arlington, VA.	
10.8 Kostelnick's Informal Tone	364	Written by Charles Kostelnick and reprinted with permission from the Society for Technical Communication, Arlington, VA.	
10.9 Informal Tone without Graph Clutter	366	Kostelnick (1998)	S-Plus with PowerPoint background
10.10 Adding Glitz with Labels	368		S-Plus with clip art

REFERENCES

Association of Research Libraries. 2002. *ARL Notebook 2002 LibQUAL+(tm) Results*. Association of Research Libraries, Washington, DC. Retrieved January 20, 2003 from *http://www.libqual.org/documents/admin/ARLNotebook111.pdf*.

Becker, Richard and William S. Cleveland. 1996. *S-Plus Trellis Graphics User's Manual.* Mathsoft, Inc., Seattle, WA, and Bell Labs, Murray Hill, NJ.

Becker, Richard A., John M. Chambers, and Allan R. Wilks. 1988. *The New S Language: A Programming Environment for Data Analysis and Graphics.* Wadsworth and Brooks/Cole, Pacific Grove, CA.

Best, Joel. 2001. *Damned Lies and Statistics: Untangling Numbers from the Media, Politicians, and Activists.* University of California Press, Berkeley, CA.

Bigwood, Sally and Melissa Spore. 2003. *Presenting Numbers, Tables, and Charts*. Oxford University Press, Oxford.

Board of Governors of the Federal Reserve System (U.S.). 1989. *Historical Chart Book*. U.S. Government Printing Office, Washington, DC.

Creating More Effective Graphs, by Naomi B. Robbins
ISBN 0-471-27402-X Copyright © 2005 John Wiley & Sons, Inc.

Brown, Judith R., Rae Earnshaw, Mikail Jern, and John Vince. 1995. *Visualization: Using Computer Graphics to Explore Data and Present Information.* Wiley, New York, p. 55.

Calle, E. E., M. J. Thun, J. M. Petrelli, C. Rodriguez, and C. W. Heath, Jr. 1999. Body-Mass Index and Mortality in a Prospective Cohort of U.S. Adults. *New England Journal of Medicine* 341: 1097–1105.

Carr, D. B. 1993. Constructing Legends for Classed Choropleth Maps. *Statistical Computing and Statistical Graphics Newsletter* 4: 15–18.

Carr, D. B. and S. M. Pierson. 1996. Emphasizing Statistical Summaries and Showing Spatial Context with Micromaps. *Statistical Computing and Statistical Graphics Newsletter* 7: 16–23.

Cleveland, William S. 1984. Graphical Methods for Data Presentation: Full Scale Breaks, Dot Charts, and Multibased Logging. *American Statistician* 38: 270–280.

Cleveland, William S. 1985. *The Elements of Graphing Data.* Wadsworth, Monterey, CA.

Cleveland, William S. 1993. *Visualizing Data.* Hobart Press, Summit, NJ.

Cleveland, William S. 1994. *The Elements of Graphing Data*, rev. ed. Hobart Press, Summit, NJ.

Cleveland, William S. and Robert McGill. 1984. Graphical Perception: Theory, Experimentation, and Application to the Development of Graphical Methods. *Journal of the American Statistical Association* 79: 531–554.

Cleveland, William S., Marilyn E. McGill, and Robert McGill. 1988. The Shape Parameter of a Two-Variable Graph. *Journal of the American Statistical Association* 83: 289–300.

Cleveland, William S. and Irma J. Terpenning. 1982. Graphical Methods for Seasonal Adjustment. *Journal of the American Statistical Association* 77: 52–62.

Downing, Douglas A. and Jeffrey Clark. 1996. *Forgotten Statistics: A Self-Teaching Refresher Course.* Barron's Educational Series, Hauppauge, NY.

Emerson, J. W. 1998. Mosaic Displays in S-PLUS: A General Implementation and a Case Study. *Statistical Computing and Graphics Newsletter* 9: 17–23.

Fisher, R. A. 1936. The Use of Multiple Measurements in Taxonomic Problems. *Annals of Eugenics*, 7 Part II: 179–188.

Fisher, Ronald A. 1971. *The Design of Experiments*, 9th ed. Hafner, New York.

Flaster, Edith. 1999. Private communication.

Hager, Peter and Howard Scheiber. 1997. *Designing & Delivering: Scientific, Technical, and Managerial Presentations*. Wiley, New York.

Harris, Robert L. 1996. *Information Graphics: A Comprehensive Illustrated Reference.* Management Graphics, Atlanta.

Heiberger, Richard M. and Burt Holland. 2004. *Statistical Analysis and Data Display: An Intermediate Course with Examples in S-Plus, R, and SAS*. Springer, New York.

Huff, Darrell. 1954. *How to Lie with Statistics.* W.W. Norton, New York.

Ihaka, Ross and Robert Gentleman. 1996. R: A Language for Data Analysis and Graphics. *Journal of Computational and Graphical Statistics* 5: 299–314.

Kosslyn, Stephen M. 1994. *Elements of Graph Design.* W.H. Freeman, New York.

Kostelnick, Charles. 1998. Conflicting Standards in Data Displays: Following, Flouting, and Reconciling Them. *Technical Communication* 45: 473 - 482.

Meyer, Eric. 1997. *Designing Infographics.* Hayden, Indianapolis, IN.

Monterey Bay Aquarium. 2001. Personal communication.

Murrell, Paul 2005. *R graphics.* Chapman Hall/CRC, Boca. Raton, FL.

Nightingale, Florence. 1858. *Notes on Matters Affecting the Health, Efficiency and Hospital Administration of the British Army*; quoted in *http://www.scottlan.edu/lriddle/women/nightpiechart.htm*.

Playfair, William. 1786. *The Commercial and Political Atlas.* William Playfair, London.

Playfair, William. 1801. *Statistical Breviary.* William Playfair, London.

R Development Core Team. 2003. *R: A Language and Environment for Statistical Computing.* R Foundation for Statistical Computing, Vienna.

Robbins, Naomi B. 1999. Creating More Effective Graphs: Trellis Display, *Proceedings of the 46th Annual Conference*, Society of Technical Communication, Arlington, VA.

Rosenfeld, Sherman. 1982. A Naturalistic Study of Visitors at an Interactive Mini-zoo. *Curator* 25: 187 - 212.

Schriver, Karen A. 1997. *Dynamics in Document Design: Creating Texts for Readers*. Wiley, New York, p. 453.

Serrell, Beverly. 1998. *Paying Attention: Visitors and Museum Exhibitions.* American Association of Museums, Washington, DC.

Spielman, Howard. 2003. Improving Business Performance through Effective Graphic Presentation. NetSession, January 30, 2003, sponsored by Applix.

State of New Jersey, Division of State Police, Uniform Crime Reporting Unit. 1994. *Uniform Crime Report: State of New Jersey*, 1994.

State of New Jersey, Division of State Police, Uniform Crime Reporting Unit. 1997. *Uniform Crime Report: State of New Jersey*, 1997.

Stevens, S. S. 1975. *Psychophysics*. Wiley, New York (from Cleveland, 1985.)

Sutcliffe, Andrea. 1996. *Numbers: How Many, How Long, How Far, How Much*. Stonesong Press, New York.

Symanzik, Jürgen and Daniel B. Carr. submitted for review. Linked Micromap Plots for the Display of Geographically Referenced Statistical Data.

Texas State Auditor's Office. 1995. Data Analysis: Displaying Data—Deception with Graphs. *Methodology Manual*, rev. 5/95.

Tisdal, Carey. 2002. Projecting Attendance: Organizational Patterns. Presentation delivered at the Visitors Studies Association Annual Conference, August 2002, Cody, WY.

Tracey, Marc. 2004. Private communication.

Tufte, Edward. 1983. *The Visual Display of Quantitative Information*. Graphics Press, Cheshire, CT (2nd ed., 2001).

Tufte, Edward. 2001. *Presenting Data and Information*. One-day course, September 17, 2001, New York.

Tukey, John. 1977. *Exploratory Data Analysis*. Addison-Wesley, Reading, MA.

U.S. Department of Energy. 1986. *Annual Energy Review*. U.S. Government Printing Office, Washington, DC.

Wainer, Howard. 1997. *Visual Revelations: Graphical Tales of Fate and Deception from Napoleon Bonaparte to Ross Perot.* Copernicus, New York. Reprinted by Lawrence Erlbaum, Mahwah, NJ, July 2000.

Walkenbach, John. 2003. *Excel Charts*. Wiley, New York.

Wurman, Saul. 1999. *Understanding USA.* TED Conferences, Newport, RI.

Zelazny, Gene. 1996. *Say It with Charts.* McGraw-Hill, New York.

INDEX

Creating More Effective Graphs, by Naomi B. Robbins
ISBN 0-471-27402-X

Copyright © 2005 John Wiley & Sons, Inc.